THE RED
HOURGLASS

GORDON GRICE

THE RED HOURGLASS

Lives of the Predators

ALLEN LANE
THE PENGUIN PRESS

ALLEN LANE
THE PENGUIN PRESS

Published by the Penguin Group
Penguin Books Ltd, 27 Wrights Lane, London w8 5tz, England
Penguin Putnam Inc., 375 Hudson Street, New York, New York 10014, USA
Penguin Books Australia Ltd, Ringwood, Victoria, Australia
Penguin Books Canada Ltd, 10 Alcorn Avenue, Toronto, Ontario, Canada m4v 3b2
Penguin Books (NZ) Ltd, Private Bag 102902, NSMC, Auckland, New Zealand

Penguin Books Ltd, Registered Offices: Harmondsworth, Middlesex, England

First published in the USA by Delacorte Press 1998
First published in Great Britain by Allen Lane The Penguin Press 1998
1 3 5 7 9 10 8 6 4 2

Printed in England by Clays Ltd, St Ives plc

For Tracy, Parker, and Abilene

Many people shared experiences and expertise with me as I researched this book; I thank them all. I'm especially grateful to David Grice, Michael Ridgway, and Michael Gabriel, whose experiences were particularly helpful; Steve Ridgway, who got me into many of these adventures; Pat Krause, who generously supplied me with information on Dr. Allan Blair; William Harrison, who shaped my first attempts to write about the black widow; Elyse Cheney, my agent, who came up with the idea for this book; and James Twiggs, who gives me my best advice.

CONTENTS

BLACK WIDOW
page 1

MANTID
page 61

RATTLESNAKE
page 85

TARANTULA
page 147

PIG
page 175

CANID
page 203

RECLUSE
page 235

BLACK WIDOW

I hunt black widow spiders. When I find one, I capture it. I have found them in discarded car wheels and under railroad ties. I have found them in house foundations and cellars, in automotive shops and tool-sheds, against fences and in cinder block walls. As a boy I used to lift the iron lids that guarded underground water meters, and there in the darkness of the meter wells I would often see something round as a flensed human skull, glinting like chipped obsidian, scarred with a pair of crimson triangles that touched each other to form an hourglass: the widow as she looks in shadow. A quick stir with a stick would trap her for a few seconds in her own web, long enough for me to catch her in a jar.

When I walk the paved paths in a certain landscaped

park in my hometown, a hot day will sometimes show me a sparkle that vanishes with any slight change of angle, and near it some windblown garbage will be lodged in the crags of a piece of granite or in the sandy dirt gathered by a prickly pear. A minute's investigation reveals that garbage, stone, cactus, and earth are all held together by an almost invisible web, at the corner of which the clawed tips of a black widow's sleek legs protrude from some crevice. To catch a widow in this situation, I have to find a live insect and toss it into her web. Only after she has come out to kill the insect and is lost in the business of biting and wrapping do I have a good chance of catching her; otherwise, she is too quick to retreat to her hiding place.

In the dry Oklahoma Panhandle, I found one under the threshold of my back door. It thrust its forelegs into the kitchen to threaten the pencil I prodded it with. Years later, when I lived in the humid Ozark Mountains of Arkansas, my wife and I had taken a new apartment, and a second before Tracy sat down on our new threshold I recognized those black lines, which might have been cracks in the cement, as a widow's legs: I yanked the spider out and captured it in a coffee can.

I have found widows on playground equipment, in a hospital, in the lair of a rattlesnake, and once on the bottom of the lawn chair I was sitting in as I looked at some widows I had captured elsewhere that day.

Sometimes I raise a generation or two in captivity. The egg sacs contain multitudes of pinpoint cannibals. Growing for several days on the residual energy of the egg yolk they consumed before hatching, they molt before ever eating. The mass of them appears as a dirty cloud at the center of the egg sac, gradually expanding into a visibly moving stain that fills the sac. They live in their private sphere for about five days before they venture out into the world through a single, perfectly round hole chewed by one precocious sister, and as they leave they trail fine silk that gleams with the sun, the group of them producing a glimmering tangle like a model of an electron cloud, the empty sac its nucleus. After a day in that tight formation, they drift away from each other.

They grow rapidly, the most successful eaters shucking a skin every few days. They begin as swirls of light brown and cream, then darken with each molt, resolving into brown with white spots. A white hourglass is soon clear on the belly. In the females, a pale orange hue dawns in the center of the hourglass with succeeding molts; the brown rapidly darkens. The orange deepens to red, like a sunset, and spreads outward to infect the entire hourglass. As adults their black is broken only by the crimson hourglass and, depending on the individual, perhaps a few other specks or stripes of red or a white dot. The male may retain his infant colors, or he may grow black and sport a psychedelic array of red, gold, and white marks.

I separate the siblings before they mature, usually when three or four remain from the original cannibal brood. It's not chance that causes these few to survive. From the beginning they were bigger, stronger, more aggressive than their sibs, and grew faster. Wild widows eat nothing for the first few days except each other; even in captivity, given plenty of small insects, the spiders prefer the taste of their sibs. The teeming masses of humbler spiderlings exist to feed the voracious few. Now I feed these few on bigger game, starting with houseflies and mosquitoes and progressing to larger insects such as crickets and bumblebees.

But I don't dare open the container until they have done their culling.

––––––––

Once, I let eleven egg sacs hatch out in a container about eighteen inches on a side, a tight wooden box with a sliding glass top. As I tried to move the box one day, I tripped. The lid slid off and I fell, hands first, into the mass of young widows. Most were still translucent cream-and-brown newborns. A few of the females were bigger and darker, but not yet black. Tangles of broken web clung to my forearms, and the spiderlings moved among my arm hairs like trickling water.

Most of them were surely too small in the jaw to

puncture my skin, but they had their toxin. The poison is there from the beginning. In the old days of the American West, Gosiute warriors ground the eggs onto their arrowheads to make them deadly.

I walked out into the open air and raised my arms into the stiff wind. The widows answered the wind with new strands of web and drifted away, their bodies gold in the afternoon sun. In about ten minutes my arms carried nothing but old web and the husks of spiderlings eaten by their sibs.

I have never been bitten.

––––––––

The black widow has an ugly web. The orb weavers make those seemingly delicate nets that poets have traditionally used as symbols of imagination, order, and perfection. The sheet-web spiders weave crisp linens on grass and bushes. But the widow makes messy-looking tangles in the corners and bends of things and under logs and debris. Often the web is littered with leaves. Beneath it lie the husks of insect prey cut loose and dropped, their antennae stiff as gargoyle horns; on them and the surrounding ground are splashes of the spider's white urine, which looks like bird guano and smells of ammonia even at a distance of several feet. This fetid material draws scavengers—ants, crickets, roaches, and so on—which become tangled in vertical strands of silk reaching from the ground to the main body of the web.

Sometimes these vertical strands break and recoil, hoisting the new prey as if on a bungee cord. The widow comes down and, with a bicycling of the hind pair of legs, throws gummy silk onto the victim.

When the prey is seriously tangled but still struggling, the widow cautiously descends and bites the creature, usually on a leg joint. This bite pumps neurotoxin into the victim, paralyzing it; it remains alive but immobile for what follows. As the creature's struggles diminish, the widow delivers a series of bites, injecting digestive fluids. Finally she will settle down to suck the liquefied innards out of the prey, changing position two or three times to get it all.

Before the eating begins, and sometimes before the slow venom quiets the victim, the widow usually moves the meal higher into the web. She attaches some line to the prey with a leg-bicycling toss, moves up the vertical web-strand that originally snagged the prey, crosses a diagonal strand upward into the cross-hatched main body of the web, and here secures the line. Then she hauls on the attached line to raise the prey so that its struggles cause it to touch other strands. She has effectively moved a load with block and tackle. The operation occurs in three dimensions—as opposed to the essentially two-dimensional operations of the familiar orb weavers.

You can't watch the widow in this activity very long without realizing that its web is not a mess at all, but an efficient machine. It allows complicated uses of lever-age, and also, because of its complexity of connections, lets the spider feel a disturbance anywhere in the web—usually with enough accuracy to tell at a distance the difference between a raindrop or leaf and viable prey. The web is also constructed in a certain relationship to movements of air, so that flying insects are drawn into it. This fact partly explains why widow webs are so often found in the facedown side of discarded car wheels—the wheel is essentially a vault of still air that protects the web, but the central hole at the top allows airborne insects to fall in. A clumsy flying insect, such as a June beetle, is especially vulnerable to this trap.

A widow's silk is strong, even by the steel-surpassing standards of the spider family. I once saw a stand of grass that seemed peculiarly matted together. It was full of wind-borne milkweed seeds and specks of chaff. As I came closer, I saw half-a-dozen grasshoppers lying at odd angles in the grass, their legs bent at painful angles and their feet generally not touching any surface. They looked as if they had been stuck there with dots of glue. I understood that I must be looking at a spiderweb, but I still couldn't see it. Only after I had looked from several different angles did the scene resolve itself into a rational arrangement, like an M. C. Escher painting in reverse.

A couple of the grasshoppers were dead and dry. Three were struggling, their great hind legs rocking and straining against the web. Another was not moving. I stared at him in the shadows of the grass, and soon I discerned a widow perched on the "knee" of his hind leg like an enormously ripe black blister. She was perfectly still. After a second she relinquished her hold on this victim and climbed toward another.

The web was shivering with the struggles of the three remaining grasshoppers. The widow visited each of them, her combed hind legs hurling silk at them in almost invisibly fine strands. The largest hopper, about three inches not counting appendages, had a charcoal-gray body with red marks. It had almost kicked free. A bubble of brown fluid—"tobacco juice," my childhood friends and I used to call it—shimmered between its mandibles and then smeared messily along a strand of web. The widow circled this big one, made a half-hearted toss of silk, and returned to the less formidable hoppers. The big one made a sustained and strenuous flurry of kicks. The widow paused, apparently gripping the web with all her legs. The web did not relent, but the hopper's right hind leg wrenched loose, and he fell free.

The others did not escape. One of them also managed to tear off his own leg, but both leg and body remained snared. As I walked away, flights of grasshoppers burst from other clumps of grass at my approach, swinging whichever way the wind carried them. The ebb of such a flight must have provided the widow's windfall.

Widows adapt their webs to the opportunities of their neighborhoods. Some choose building sites according to indigenous smells. Webs turn up, for example, in piles of trash and rotting wood, the web holding together a camouflage of leaves, dirt, or bark. A few decades ago the widow was notorious for building its home in another odorous habitat—outdoor toilets. Until the 1970s the outhouse was the source of the majority of black widow bites in the United States. (The outhouse is still the likeliest place to get bitten in parts of Africa and Australia.) People were most often bitten on the genitals or buttocks, with men suffering more bites than women for the obvious anatomical reasons. The widow would make a web under the seat or across the hole of the toilet. The insect-attracting odors, combined with a design that funneled insects through a single entrance, provided ample prey, and the darkness of the setting pleased the widow's sensibilities.

When part of a human body intruded on the web, the widow would often treat it as prey and bite it. In other cases, the human intruder would crush the widow, provoking a defensive bite. My grandmother, who grew up in rural Oklahoma and lived in a dugout and in other houses without plumbing in her youth, told me that some people would habitually take a stick to the outhouse with them. Before conducting the business for which they had come, they would scrape under the seat and inside the hole with the stick, listening carefully for a sound like the crackling of paper in fire. This sound is unique to the widow's powerful web. Anybody with a little experience can tell a widow's work from another spider's by ear.

Man-made objects provide so many inviting angles and crevices for the widow that scientists consider it commensal to us, a generally harmless user of our inadvertent services.

Widows have colonized several islands by hitching rides in human vehicles. The South Atlantic island of Tristan da Cunha was considered free of widows until a botanical survey discovered some in 1968. The United States Coast and Geodetic Survey had brought satellite-tracking equipment to the island. The equipment had been used in Australia, which must be where red-back spiders climbed aboard. *Red-back* is an Australian and Southeast Asian name for the black widow; the species found there has an extra red stripe like a bursting seam along its back. Widows can travel with large, aggressive animals armed with flyswatters because they have a talent for going unseen. They are so widespread, and are mentioned in the oral traditions of so many peoples, that they clearly did not wait for human help to conquer the world, but hitchhiked with us when they got the chance.

The widow is found in virtually every temperate or

tropical place in the world, under such aliases as *cherry spider, black wolf, twenty-four-hour spider, night stinger, shoe button spider, coal-black lady,* and many others. The name *black widow* is only about a hundred years old in English, but similar names in Italian and Russian are much older. Twenty or thirty species of widow spiders exist, varying in color and other minor details. The United States has a red widow and a brownish-gray one, as well as three black species. But all of these species around the world have similar lifestyles, and all are dangerously toxic. Scientists group them in the genus *Latrodectus,* a name that translates as something like "sneaky biter." The roots of that term go back to ancient Greece, where it was a name for vicious dogs.

———

Widows move around in their webs almost blind, yet they never make a misstep or get lost. In fact a widow knocked loose from its web does not seem confused; it will quickly climb back to its habitual resting place. Furthermore, widows never snare themselves, even though every strand of the web, except for the scaffolding, is a potential trap. A widow will spend a few minutes every day coating the clawed tips of its legs with the oil that lets it walk the sticky strands. It secretes the oil from its mouth, coating its legs like a cat cleaning its paws.

The human mind cannot grasp the complex func-

tions of the web, but must infer them. The widow con-
structs it by instinct. A ganglion smaller than a pin-
head—it's too primitive to be called a brain—contains
the blueprints, precognitive memories the widow un-
folds out of itself into actuality. I have never dissected
with enough precision or delicacy to get a good specimen of the black widow's tiny ganglion, but I did
glimpse one once. A widow was struggling to wrap a
mantid when the insect's forelegs, like scalpels mounted
on lightning, sliced away the spider's carapace and left
exposed among the ooze of torn venom sacs a clear
droplet of bloody primitive brain.

———

The cold was coming on. You could see it in the masses
of black flies congregating on the wound I'd made by
pruning an elm: the sap still ran with the heat, but the
flies were slow enough to pick up with your fingers,
slow with the premonition of a freezing still two days
distant. Moths flocked at the marigolds in shivers of
gray and brown; green bees crawled the flowers, their
legs thickening with pollen; green and black grasshoppers the size of human fingers knocked against the
fences; pale red ants swarmed the fading grass around
their sandy hills, their motion frantic but much slower
than during the summer mating flights. Everything was
in the state it maintains for perhaps three days in the
autumn, that state of reckless swarm and copulation

and feeding slowed by the death already unfolding in the insects' ephemeral bodies.

One day I came to see a black widow I'd been watching for a month or so. Her home was in a wooden gate. The web stretched across the front of it and into its interstices and crevices. Most of the silk was hidden between boards. I hadn't seen the widow for a few days. She had probably mated and crawled into some tight space near the ground to winter; she would lay her eggs in the spring. I began to tear down the web, using a stick in case I was wrong about her departure, and the sound was like a staticky radio. As I dragged the stick through cracks in the gate, desiccated prey fell out on the sidewalk. I examined the webbed carcasses of beetles, gnats, lacewings, ants, and a male black widow. Then I knocked loose something unusual.

It was a paper nest. The brown-and-yellow wasps that build such nests are social. The queen wasp builds the first two or three cells—each a hexagonal cylinder hanging with the open end down—of chewed wood pulp and saliva, deposits eggs in them, and feeds the larvae that hatch with chewed-up insects and spiders. I once stole a nest of larvae. Their wormlike bodies were white, almost iridescent, and their heads were hard brown orbs with two dark nubs of undeveloped eyes and four fingerlike mouthparts waving hungrily. I fed them drops of milk-and-sugar from the tip of a sewing needle. They sucked it down hungrily and then waved their heads around, mouthing for more. Sometimes the drop would fall off prematurely and mantle their white "shoulders." They would wrench their heads in big arcs, futilely attempting to get at the milk. They looked disturbingly like kittens do before their eyes open.

I kept them alive this way for almost two weeks. They should have been ready to pupate, but I had no way to stimulate the hormonal changes that would have brought them to that state. In the wild, the queen or older sibs would have fed them the secretions of their own bodies that would not only transform them into adults but determine their gender—male, female, or the sterile, pseudofemale form of the worker wasp.

As a first generation of workers enters the pupal stage, the queen seals the cells with a papery cap. The adult wasp's shape emerges slowly during pupation. (Imagine a figure carved in wax slowly melting into a shapeless mass—but imagine it happening backwards.) The process can be observed by tearing open older nests in which several generations of wasps are at different stages of pupation. The ones nearly ready to wake are hard and their colors vague; they look like mummified adults folded into fetal position.

The first generation of workers emerges from pupation to take on the duties of guarding the nest, gathering food, and caring for younger sibs. The queen becomes a specialist in egg-laying and does little else.

A worker stings by shoving its specialized ovipositor

into the enemy's flesh. It injects an acidic toxin that kills most insect-sized enemies quickly. To a human, the sting feels like a burn from an inextinguishable match imbedded in the skin.

The wasps that live through the winter are a new generation of queens, reared through the larval stage before the cold hits and left to pupate. The workers and the old queen die off gradually, lingering in the gathering cold, moving in slow motion as parts of their bodies freeze, blacken, and wither. In the spring the new queens chew out of their cells to disperse and found new nests, or sometimes to stay in the same nest. There, one conniver gains dominance by secretly eating her sisters' eggs. It's like a Bette Davis movie in miniature.

The nest that fell out of the black widow's web had died at the moment of awakening. That much was clear because a dozen adult queen wasps lay dead in sealed cells, while one was dead in mid-emergence. This one had chewed her way out of her cell. Her next step should have been to crawl into sunlight, spreading her wings to gather the energizing heat. But while the wasps lay dormant in the early spring, the black widow chose the gate for a web site. She had built her web across the wasp nest, anchoring many strands on its surface. In fact, she had covered the nest so tightly that most of the wasps couldn't emerge. Several of them had chewed away the paper caps of their cells only to be held in by the strands of web. The widow must have detected the one wasp on the edge of the paper nest emerging. The wasp was webbed up, half out of the cell, but with her wings and stinger still trapped inside. The tough silk looped around her antennae like a lariat around a longhorn steer's rack. The widow had probably killed her with a bite on the joint where the antenna bends—this is the widow's favorite way to attack wasps, suggesting that she knows about the danger of the stinger. This wasp was structurally undamaged, but pale and hollow: when I held her to the light, the light passed through.

The rest of the wasps had died in their cells, either starved or eaten alive. It wasn't the widow that had eaten them; it was some band of scavengers. Unlike their emergent compeer, these wasps had mostly been gnawed from the stinger down toward the head. Something had started from the base of the nest and eaten the paper, moving down into the cells. The unknown scavengers apparently made no distinction between paper and wasps, for some of the wasp bodies had been eaten—some of them only halfway. In a few cells the scavengers had reached the wasps' heads, devoured them, and gnawed through the paper caps below. The tiny, round openings the scavengers made were easy to distinguish from the sheared holes made by a wasp's mandibles.

Some of the scavengers had crawled out onto the lower surface of the wasp nest—a bad move, as it turned out. The widow had killed a good number of them. Unlike the wasp's, these creatures' exoskeletons did not remain intact once their moisture had been siphoned out. Their remains were scattered fragments of chitin I couldn't readily identify.

I called an entomologist for help. He mentioned parasitic wasps that devour larger wasps as they hibernate. These clearly were not the culprits here, because my vandals had been just as interested in paper as in protein. The entomologist also suggested a few scavenging varieties of beetles and caterpillars. Some tiny beetles, for example, make a habit of ruining insect collections.

The beetle hypothesis matched the remains I had found. As I looked at the mysterious fragments of chitin again, I saw that their shapes could easily be the hollowed-out abdomens and thoraxes and wing cases of beetles. The question seemed settled. Up to this point I had tried not to damage the wasp nest too much, since I might thereby miss some detail that would help me solve the mystery, but now I went about dismantling it. As I extracted the wasp carcasses, I noticed a whitish gleam at the tops of their cells. I cut away a few cell walls with scissors. I poked into the whitish mass. Was it caterpillar silk? If so, I would have to change my beetle hypothesis.

Suddenly a white spider with brown spots emerged and went waddling across my hand. Its tiny body was almost all abdomen, a soft abdomen with textured bands that resembled rolls of fat. It wasn't a black widow. It was a member of a common species I had seen many times before, in the corners behind furniture or under pieces of siding, a creature so unobtrusive it has no common name.

After capturing the little spider, I reinspected the silk on the wasp nest. The silk on the surface was, as I had assumed, the tough fiber of the black widow, and the widow had definitely scored the emergent wasp and some of the scavenging beetles. But inside many of the cells were tiny snares made of softer silk, the delicate work of the little white spider. Thanks to the tunneling of those paper-eating scavengers, a cavity joined most of the cells at the top, and in this cavity the white spider had been living. His web held the remains of dozens of the beetles. It was an ideal arrangement for the two spiders, the smaller protected from the larger by the remains of the wasp nest, both feeding on what must have been a great wealth of prey—a symbiosis between two predators, each presumably unaware of the other.

The widow had, by its choice of web site, exterminated a half-dozen queen wasps who might have produced nests of their own and who should at this season be dying off, having seen larval daughters into pupation.

I found a similar case of miniature genocide in a widow's web built over the egg case of a mantid. The egg case was about the length of a large paper clip—an oval mound of beige lying unobtrusively on the piece of wood to which it was attached. A single case can hold two hundred eggs. The young mantids had emerged from the egg case, and dozens of small, thready knots in the widow web showed what had happened to them.

———

Widows have been known to snare and eat mice, frogs, snails, tarantulas, lizards, snakes—almost anything that wanders into that remarkable web. I have never witnessed a widow performing a gustatory act of that magnitude, but I have seen them eat scarab beetles heavy as pecans, cockroaches more than an inch long, bumblebees, camel crickets, and hundreds of other arthropods of various sizes. I have seen widows eat butterflies and ants that most spiders reject on the grounds of bad flavor. I have seen them conquer spider-eating insects such as adult mantids and mud dauber wasps. The combination of web and venom enables widows to overcome predators whose size and strength would otherwise overwhelm them.

Among the widow's more interesting habitual enemies is a certain carabid, or ground, beetle. There are thousands of species of carabids, but the one I'm talking about runs about the size of a domino, with mandibles over a quarter of an inch. A pair of these serrated mandibles resembles the claw of a crab. They pinch shut with unbelievable force.

I became interested in these beetles one particularly wet May when an abundance of earthworms writhed all over the sidewalks and the grass. The worms congregated by the dozens under rocks and lawn furniture, a few of them hobbled in their conjugations, the rest slithering away into the wet earth at any intrusion. Looking at earthworms in the loose upper layer of soil, stirring among their castings, I disturbed many carabid beetles. They would storm off when I uncovered them, somehow thrusting the substantial bulk of their black bodies, gleaming with the dampness of the earth, into the ground within a second or two of being exposed. When I tried to dig them out, they had already vanished, as if they'd converted themselves into bits of the fertile earth they lived in.

Their mandibles marked them as predators. I looked for them and found them abundant: here a carabid dismantling the grub of a June beetle, its mandibles cutting the grub as easily as scissors would; there another snipping an earthworm in half. I captured one carabid by harassing him with a stick. He seized the stick between his mandibles and did not let go until after I had lifted him into a jar.

His predatory habits were spectacular. He would attack any moving thing immediately. I offered adult June beetles. The carabid would rush one, seize it, and work his mandibles over the June beetle's body until he had a grip on the juncture of abdomen and thorax.

Then he would squeeze until the June beetle broke in half with a loud crack. He would lap the juice out of the abdomen as the head and thorax of the dismembered prey crawled away in a panic.

The carabid's next victim was a tomato hornworm, which is actually the caterpillar of a gigantic sphinx moth. The caterpillar was about the size of my middle finger. I removed it from a tomato plant in our garden and placed it on the lip of a two-gallon can. It rippled around, gripping the brim of the can, making a complete circle in a few seconds. It did not stop.

I left it and returned two hours later to find the creature still circling, its pace and path unaltered. The huge green caterpillar might have crept endlessly in its circle. Picking it up gently, I set it back on the lip pointing in the opposite direction. It circled, and circled again. . . .

I decided the caterpillar was too stupid to live. I put it into the carabid beetle's container. The caterpillar was much larger, but it had no means of defense. The carabid sliced into it and lapped at its leaking blood. Because the caterpillar was so big, the carabid had to repeat his attack eight or ten times. The caterpillar crawled away frantically for the first few wounds, but it was so slow that its movements hardly inconvenienced the beetle drinking from its bleeding flank. After ten minutes or so the caterpillar lay still. Its jade flesh turned black as the beetle chewed and drained it. After half an hour the entire body was a black heap about a quarter of its original size. It lay in the dirt like an empty burlap sack. The beetle stood with his head raised and his mandibles flexing. He looked something like a bellowing bull and might have been humorous if he hadn't just committed an awesome display of predation.

The carabid was insatiable, and I eventually offered him a great variety of prey. He tried to eat a small toad but couldn't get a good enough grip to kill it. Once I put the carabid in a jar with a large gray wolf spider. The spider was missing a leg because I had injured it in the capture. The next morning the carabid was circling the jar looking for his next meal; all that remained of the big spider were seven gray legs.

In late autumn, as the supply of prey was running out, I realized I would have to sacrifice either the carabid or one of the widows I was also keeping. The choice was not mine; I could only put the carabid in with a widow and see which fed and which died.

The fight, if it can be called that, was over in about three minutes. The heavy carabid was half a foot above ground, arching his body against the gummy strands that had hoisted him. His mandibles slashed and scissored at the web, doing no damage at all. The widow circled just out of range of his mandibles and his kicking legs, picking her chances to hurl silk strategically. Soon she had his mandibles roped shut. Reaching delicately past that awesome and now useless set of hardware, she bit him on an antenna. He thrashed a minute longer, and then was food enough to last her through the winter.

Since then I have noticed the remains of carabid beetles in or beneath widow webs many times. The hard black exoskeletons seem immune to erosion and decay; they lie in piles of rot for months, maybe years, without losing their striking luster. Once I removed the head of a carabid, now hollow and dry, from a widow web and found two narrow strips of what looked like transparent tape projecting from the rear of the head. When I tugged on these, I realized they were the tendons that controlled the mandibles. I used them like puppet strings to make the disembodied head bite. It would pick up pencils, twigs, and bits of gravel this way. I even put my little finger between the mandibles and caused the beetle head to bite me, but it wasn't painful. I couldn't generate nearly as much force as a live carabid can, with the tendons anchored far back in the thorax.

———

The widow routinely knocks off larger predators, but, like every other animal in the world, it sometimes serves as an entrée for something else.

The mantid is a unique danger because of its unusual weaponry, but even this superpredator doesn't always survive the widow. Widows are sometimes paralyzed by mud daubers and other wasps, who use them as live food for their larval young. The innocuous-looking daddy longlegs spider and some of its kin are said to eat widows, as are certain lizards.

A century or so ago, when black widows and various other small and mysterious predators had been insufficiently studied and most people had only the exaggerations of folklore to go on, people would stage fights between such predators, creating a Colosseum spectacle in miniature. The participants included Gila monsters, widows, tarantulas, scorpions, small rattlesnakes, and mantids. One can find similar activities mentioned in histories and travelogues from various cultures. Tarantula fighting is supposedly still common in the Philippines, and the Chinese had an elaborate system for the sport of mantid fighting.

The battles between the mini-superpredators in the United States were generally staged for gambling purposes or as advertising gimmicks (one such fight went on in the window of a general store). We know about them mostly from newspaper stories, which covered them as curiosities or sporting events; I've even seen a paper from an Old West town in which a bug fight was the lead story on page one. The fights could last days— or, if mutually uninterested combatants were chosen, past the tolerance of the observers. Outcomes varied according to the method of staging, but the widow, smallest in the field of competitors, fared respectably. Even rattlesnakes proved vulnerable to the widow's venom.

While such spectacles no longer meet the average editor's requirements for serious journalism, people still stage them for gambling or just for their private amusement. I came across a reference to this practice in a 1993 article in *Harper's Magazine*. The article is about a man who stole thousands of rare books from libraries, but it mentions in passing that he and a traveling companion, while stopping in Amarillo, Texas, tried unsuccessfully to make a widow and a tarantula fight in a coffee can. The book thief's mistake was in not giving either spider an environment suited to its hunting methods. Since this diversion is pretty much irrelevant to the article's main subject, I suppose it was included to show just how strange the book thief was, or perhaps to show how people behave in Amarillo.

Another participant in these contests was the windscorpion. This creature is known, where it is known at all, by many names: *solpugid* ("sundagger"), *sunspider,* and *matavenado* (Spanish for "deer-killer," though it does no such thing). It is actually neither a scorpion nor a spider. It constitutes a separate family within the arachnid class that contains both. Superficially, it resembles a spider, and its hunting habits are similar to those of wolf spiders—though the windscorpion can grow up to five or six inches long. But a closer look (which is hard to get: the animal is nocturnal and the fastest run-ner of all the arthropods) shows an upswept, segmented abdomen like a scorpion's. The pedipalps, the leglike feelers at the front end, are clubbed and sticky, for nab-bing prey. Two prominent eyes sit above enormous jaws, proportionately the largest found in any known animal; in one specimen I collected, the jaws constituted one-third the total body length. The windscorpion kills by the mechanical injury it inflicts with these instru-ments. It has no venom.

In the desert of the American Southwest, windscor-pions and widows thrive—both love heat. The two predators compete for the available insect prey, and they also readily devour each other. The windscorpion is an avid eater of widows. It chews them into scribbles of skewed legs and pulp, sucking down body fluids that literally are poisonous enough to kill a horse, but that have no adverse effect on the windscorpion. However, even windscorpions have a less-than-sterling record against the widow, which frequently snares and eats them. In many deserts around the world, true scorpions make this a three-way rivalry. Seven-inch scorpion husks have been found in widow webs.

A widow's most dangerous enemy is another widow. An adult female will fight any other female who crowds her, and the winner often eats the loser. I am told that staged fights between widows are still a popular enter-tainment in Mexico: children put the widows on a stick and pass it around so everyone can see. Sometimes one female ties another up and leaves without killing her. I've seen this happen several times with widows in captivity, and in the wild I once came across what looked like a black pearl wrapped in silk on a red fence. When I peeled off the silk, the pearl unfolded its legs and rushed away. Another time I saw a female widow bind another and bite her on a leg joint, just as she would do to a prey item, and then leave without feeding. Soon the beaten widow stretched her legs, shrugged off her fetters, and walked away, becoming the only arthropod I've ever seen survive a widow's bite.

———

The widow gets her name by eating her mate, though this does not always happen. When a male matures with his last molt, he abandons his sedentary web-sitting ways. He spins a little patch of silk and squeezes a drop of sperm-rich fluid onto it. Then he sucks the fluid into the knobs at the ends of his pedipalps and goes wandering in search of females. When he finds a web, he recognizes it as that of a female of the appropriate species by scent—the female's silk is laden with pheromones. Before approaching the female, the male tinkers mysteriously at the edge of her web for a while, cutting a few strands, balling up the cut silk, and otherwise altering attachments. Apparently he is sabotaging the web so the vibratory messages the female receives will be imprecise. He thus creates a blind spot in her view of the world. This tactic makes it harder for her to find and kill him. Then he's ready to approach her. He distinguishes himself from ordinary prey by playing her web like a lyre, stroking it with his front legs and vibrating his belly against the strands.

I came upon one courtship in progress. The male was brown and white. A broad white bar marked the midline of his back; the hourglass on his belly was white. He tapped his two front legs on the web before him like a blind man tapping his cane. The female was near the tubular retreat in a sheltered corner of her web. She responded by staying still. He approached and turned to bicycle his hind legs, roping her legs with fine silk. She stirred briefly, as if settling in her sleep, the strands of web he'd thrown sliding off her; he fled to the far side of her web and hung there licking the tips of his claws. The next time he approached, the female responded to his leg-tapping with a tap of her own foreleg—the same move she would make if she suspected prey at hand. Its impact on the web sent him running again.

He never gave up, but approached her time after time. Sometimes he retreated for no reason visible to me. Sometimes she seemed momentarily hypnotized by his tapping routine; other times she tensed as if she had detected prey and were "listening" to the vibrations of her web for a direction. I counted over one hundred retreats before an appointment forced me to leave. The next morning I looked at the web. The female was nowhere in sight. At the center of the web hung a small bundle of silk from which a number of translucent legs protruded. I removed the bundle carefully, the web tearing with its characteristic dry crackle. Once I had unwrapped the male's body, I found little structural damage to the exoskeleton. It had turned golden instead of brown with the loss of blood and organs. The palps were still dark and round and probably full.

No explanation for the female black widow's unpredictable response has held up for long. Sometimes she eats the male without first copulating; sometimes she snags him as he withdraws his palp from her genital pore; sometimes he leaves unharmed after mating. Scientists used to assume the female only ate the male if she had already mated with a different male, or if she was particularly hungry. But no such pattern has proved true. Recently fed virgin females sometimes eat males.

I have witnessed male and female living in apparently platonic relationships in one web. Sometimes the male's attempts to mate, and the female's attempts to run him off, last for days or weeks before some decisive ending occurs. Occasionally a male will devour a female in captivity. Whether this occurs in the wild I don't know, but such abortion of future generations obviously couldn't be very common.

Captivity may play an enormous part in our understanding of widow mating—or, rather, our misunderstanding. In the wild, a male will quickly run away after mating. A researcher named R. G. Breene found that captive males, if removed from the container of a female, could breed repeatedly, fertilizing the eggs of a whole series of females. He proposed that this outcome is the normal one, and that sexual cannibalism, while it does happen occasionally, is fairly unusual. The high incidence of mate-eating found by scientists in laboratories, says Breene, can be explained by the sealed containers used in labs. The male can't escape as he naturally would, and sooner or later he gets eaten. Breene has even suggested that John Henry Comstock, the arachnologist who popularized the name *black widow* around 1900, was observing under just such misleading conditions and used his misinformation to foist the "widow" name onto the general public. Breene suggests males eaten in the wild are generally malnourished or sick. Though he doesn't go this far, his notion would allow us to see the female's cannibalistic brand of romance as an instance of natural selection, with the female scourge culling imperfect males from the gene pool before they mate, or at least before they mate again.

Breene's position is weakened by the more-than-occasional observation of sexual cannibalism in the wild—I've seen it several times, and so have others. I used to keep a male and female of breeding age in a mustard jar together, and they never injured each other, disproving Breene's idea that the captive situation invariably produces cannibalism. And then there's the red-back.

The red-back widow of Australia enacts a startling variation on the motif of mate murder. When the red-back male has inserted the tip of his palp into the female's genital pore, he does a somersault, bringing his abdomen to the female's fangs. She bites him and begins to digest and suck out his innards while they are still copulating. He sacrifices himself, perhaps helping to ensure protein and calories enough for the female to lay eggs.

Then again, scientists have not been able to show an increase in egg-laying among females who have eaten males, which are, after all, skimpy—often the female outweighs the male by a factor of fifty. Another theory is that the males hold the females' attention longer, and are allowed to copulate longer, by offering their bodies as a sort of diversion. Their self-sacrifice reduces the mating opportunities of rival males.

None of the widows I've been around has sacrificed itself the way the red-back sometimes does, but the strategy should still work. A female whose time is taken up by an elaborate series of threats and retreats before mating doesn't have time to receive other prospective mates. A freshly mated female who's busy eating suitor number one may not want to be disturbed at her meal by suitor number two. In this view, the most genetically successful males are those who occupy the female's attention longest, even if it means sticking around to become a meal.

Mating is the last thing a male does. Once he's left his web to seek mates, he never eats again; and whether he finds females or not, he is already wasting away, collapsing toward his preordained life-limit, which is marked by the coming of the cold.

———

Many widows will eat as much as opportunity gives. One aggressive female I collected on the back porch of my parents' house had an abdomen a little bigger than a pea. She snared a huge cockroach and spent several hours subduing it, then three days consuming it. Her abdomen swelled to the size of a largish marble, its glossy black stretching to a tight red-brown. With a different widow, I decided to see whether that appetite was really insatiable. I collected dozens of large crickets and grasshoppers and began to drop them into her web at a rate of one every three or four hours. After catching and consuming her tenth victim, this bloated widow fell from her web and landed on her back. She remained in this position for hours, making only feeble attempts to move. Then she died.

The widow's appetite is connected to its reproductive prowess: the more a female eats after mating, the more eggs she can lay. I rarely find a black widow in the wild with more than two or three egg sacs, but in captivity, where the spiders' work of fighting the elements and repairing the web is reduced and the prey abundant, I've seen them produce many more. Some widows lay eggs until their bellies shrivel like raisins and they die. When an egg sac is taken away from a widow, she lays a new clutch that night. As long as the female widow keeps eating, she can make more eggs, but the last clutch or two in a long series are usually sterile. The greatest number of sacs I have seen one female produce is nine, of which six hatched a normal number of spiderlings. From the seventh a small brood of a dozen hatched. The eighth and ninth sacs proved sterile.

The female starts her egg-laying ritual by spinning the beginnings of the sac—a short stem from which hangs a flat patch of webbing like a lace doily. Working belly-up, the widow lays her egg mass on the lower surface of this patch. The egg mass slowly squeezes out of the genital pore in the middle of the belly. It is a gooey substance in which no individual eggs are visible, and it comes out in a nearly perfect orb. The process resembles the blowing of a bubble with chewing gum, but can last hours. The egg mass, which is the color of butterscotch pudding or a little lighter, sticks to the silk platform.

After resting, the widow returns to work on the sac. Hanging beneath the egg mass in the usual belly-up position, she pokes the spinnerets that tip her abdomen at the edge of the platform and squirts silk at it. This silk is fine; as it comes out and adheres to the platform, you can see the individual fibers accreting into a lacy tangle. The edge of the structure always has spaces like those in lace, but the rest is solid; somehow, the silk dries into an unbroken, waterproof fabric. Its texture is like nothing else the widow spins; it is somewhere between paper and linen. As she adds to the platform, it grows to resemble an inverted goblet; then the goblet rounds in to form an orb, which does not touch the inner orb of the egg mass except at the original point of attachment. You can see the darker egg mass inside the sac by holding it in front of a light.

After the egg sac dries, it usually has a smooth surface with a single nipple at its starting point. Its color darkens a bit; the darkest become light brown or manila, and others remain almost white. Its fabric is tough; when you tear into it, you can see intricate layers within. The egg mass inside slowly dries, resolving itself into individual eggs that look like the grains of sand in an hourglass.

———

I worked as a groundskeeper at a hospital one summer when I was in college. One day I was cleaning out a cement pit that opened onto the basement laundry. The pit was full of dead leaves and discarded containers and other trash, and the vents from the laundry poured steam onto me as I swept. Whiptail lizards as long as my hand would streak out of the trash and lodge a few feet from their original hiding places. One ran across my boots. I ran after him and dug through the pile of junk where he went to ground.

I forgot the harmless reptile when I turned over a five-gallon plastic bucket and found it filled with the web of an immature female widow, who hung at the center of the bucket, the bright hourglass nearly covering the ventral surface of her slender abdomen.

She did not seem to have noticed that I had turned her world upside down. I was shocked. I had been thrusting my hands into such junk all afternoon. I aimed the handle of a rake at the spider pool-cue style and with a tap converted her to a paste on the bottom of the bucket. I felt pleased about having dispatched her so economically, but soon I regretted it. She was, after all, a good-looking specimen, and I could have enjoyed watching her with a safe layer of glass between us. I told myself I had prevented the possibility of anyone else's getting bitten, but, since no one but me would find himself cleaning out this hole in the ground in the near future, my rationalization didn't entirely convince me.

I forgot this moral dilemma when I found a second widow. And a third, and a fourth and fifth. I captured them in discarded paper cups and styrofoam Big Mac containers (this was before McDonald's became environmentally concerned). They were all immature, all colored brown and white except the one I had killed. I stashed them in some sunless corner so that I could retrieve them after work. Later, when I told my boss about seeing black widows, he said, "I hope you killed them." I changed the subject.

I took those widows home after work and installed them in terraria. As the summer passed, I found other widows at the hospital. I found them on three sides of the building; I found them low and high; I found them at various ages.

One day I was cleaning out a latticed stone wall. I suppose it was designed to throw a lovely dappled shadow on the walk, but what it actually did was provide crevices for trash to lodge in when the wind got stiff. I reached into the wall's triangular holes to fish out candy wrappers and floral-print Dixie cups that must have come from the hospital's family room. The wall was thick; my hand went in up to the middle of my forearm. In a dazzling display of intelligence, I was working bare-handed.

I thrust my hand into the next triangular space and felt my finger brush something desiccated and velvety. I pulled my hand out immediately. I had no idea what I had touched. I had never felt anything like it. I looked into the hole.

A female widow hung in a torn web with dry leaves and a few scraps of shredded paper. It was the widow's abdomen I had touched, but, because she had not re-treated at my intrusion, I knew she was dead. I pulled everything out of the hole.

The widow's exoskeleton was in good shape, but it was dry. Her leaking blood had made a sugary crust where abdomen joined cephalothorax. She had simply died, as widows do after laying their last egg sacs. That last egg sac hung in the web too. It still held its globular shape, but its crisp beige had gone a little gray with age. A single pinhole showed that the spiderlings had left. I tore the sac open. The tiny individual eggs were visible, each broken open so that its shape was ruined, a neatly halved orange rind instead of a spherical orange. These shells had dried to a sandy grit. The web also held the wrapped carcasses of a few young black widows. They had been eaten by siblings or maybe by their mother in her last languid days.

The remaining spiderlings must have dispersed soon after hatching that spring, as young widows do after their infant phase of cannibalism. After crawling out of the wall, each produced a few filaments of web, which caught the wind; when the filaments had spun out long enough, the wind lifted strand and spiderling and car-ried them as far as it wanted to take them. This strategy for dispersal is called *ballooning;* it is common in spi-ders. Human balloonists have encountered ballooning spiderlings high in the atmosphere; scientists have found spiderlings hundreds of miles from their birth sites.

In the case of the black widow, it's not unusual for the spiderlings to land very close to the mother's nest, so that a parabola of new webs springs up on the ground downwind from the point of origin. A few years of this pattern of dispersal can infest miles of open ground. The widows can exist so close to each other because, even though they're cannibals, they seldom leave their own webs. Infestations of this size crop up in grassy fields, and farmers have burned off stretches of pasture to destroy the black widows. Putting sheep or pigs on the fields is also supposed to clear them of widows. Sheep react only slightly to widow venom, probably because of their long evolutionary history of grazing low in the spidery grasses. Pigs are not immune, but a layer of subcutaneous fat tends to keep poisons out of their bloodstreams, so that they often survive the bite of a widow. Their rooting destroys widow webs. The U.S. Navy once arranged for men to clear an infested field with flamethrowers.

What I had been seeing at the hospital that summer was the dispersal of the widows in reverse. I had found the young early and then, near the end of the summer, discovered their nest of origin. Of course, some or all of the young widows I had found may have come from other nests; I'll never know. But all the females I found there that were old enough to show their adult colors had the same pattern as the dead mother—no red marks except the double triangle on the belly.

The last few weeks of the summer I often noticed the wind. It made the roof of the building creak; it sent trash scudding around corners. It eddied into odd nooks of the building to deposit dirt and dead leaves. It stopped short, dropping things from high up—pine needles, dust, scraps of paper on which were written the names of diseases and drugs, bits of newspaper, dandelion seeds on their tufted parachutes, bits of animal fur, bits of gossamer.

———

The first thing people ask when they hear about my fascination with the widow is why I am not afraid. The truth is that my fascination is rooted in fear.

I have childhood memories that partly account for my fear. When I was six my mother took my sister and me to the cellar of our farmhouse and told us to watch as she killed a widow. With great ceremony she produced a long stick (I am tempted to say a ten-foot pole) and, narrating her technique in exactly the hushed voice she used for discussing religion or sex, went to work. Her flashlight beam found a point halfway up the cement wall where two marbles hung together—one crisp white, the other a glossy black. My mother ran her stick through the dirty silver web around them, and as it tore she made us listen to the crackle. The black marble rose on thin legs to fight off the intruder. As the plump abdomen wobbled across the wall, it seemed to be constantly throwing those legs out of its path. It gave the impression of speed and frantic anger, but actually a widow's movements outside the web are slow and inefficient. My mother smashed the widow onto the stick and carried it up into the light. It was still kicking its remaining legs. She scraped it against the sidewalk, grinding it to a paste. Then she returned for the white marble: the widow's egg sac. This, too, came to an abrasive end.

My mother's purpose was to teach us how to recognize and deal with a dangerous creature we would probably encounter on the farm. But of course we also took the understanding that widows were actively malevolent, that they waited in dark places to ambush us, that they were worthy of ritual disposition, like an enemy whose death is not sufficient but must be followed with the murder of his children and the salting of his land and whose unclean remains must not touch our hands.

The odd thing is that so many people, some of whom presumably did not first encounter the widow in such an atmosphere of mystic reverence, hold the widow in awe. Various friends have told me that the widow always devours her mate, or that her bite is always fatal to humans—in fact, it rarely is, especially since the development of an antivenin. I have heard told for truth that goods imported from Asia are likely infested with widows and that women with bouffant hairdos have died of widow infestation. Any contradiction of such tales is received as if it were a proclamation of atheism.

Scientific researchers are not immune to the widow's mythic aura. The most startling contribution to the widow's mythical status I've ever encountered was *Black Widow: America's Most Poisonous Spider,* a book by Raymond W. Thorpe and Weldon D. Woodson that appeared in 1945. This book apparently enjoyed respect in scientific circles. It was cited in scientific literature for decades after it appeared; its survey of medical cases and laboratory experiments was thorough. However, between their responsible scientific observations, the authors present the widow as a lurking menace with a taste for human flesh. "Mankind must now make a unified effort toward curtailment of the greatest arachnid menace the world has ever known," they proclaim. The widow population is exploding, they announce with scant evidence, making it a danger of enormous urgency. Perhaps the most psychologically revealing passage is the authors' quotation from another writer, who said the "deadliest Communists are like the black widow spider; they conceal their *red* underneath."

We project our archetypal terrors onto the widow. It is black; it avoids the light; it is a voracious carnivore. Its red markings suggest blood. Its name, its sleek, rounded form invite a strangely sexual discomfort: the widow becomes an emblem for a man's fear of extending himself into the blood and darkness of a woman, something like the vampire of Inuit legend that takes the form of a fanged vagina.

———

The widow's venom is, of course, a soundly pragmatic reason for fear. People who live where the widow is common have known about its danger for centuries; from Russia to North America, folk wisdom carried warnings and remedies. However, the medical establishment was slow to accept the widow as a killer of humans. The creature seemed too small to be responsible for the things she was charged with—extravagant suffering, painful death. People bitten by the spider sometimes didn't link it with the symptoms that developed hours later; if they did, doctors assured them the spider was not the cause.

Virtually all spiders use some sort of toxin to subdue prey; the question arachnologists were still debating into the twentieth century was whether any of these toxins, in the small doses delivered by spiders, could harm people. Many doctors treated black widow bites and believed their patients' surmises about the source of the problem, but the larger scientific and medical community remained skeptical. The skeptics didn't find the anecdotal evidence sufficient. They wanted definitive laboratory evidence, the kind that could be replicated.

Starting in the late nineteenth century, many workers attempted to deliver such evidence in the form of animal experiments.

Reports of such animal tests—they still go on today, as scientists try to understand how the venom works—read like H. G. Wells's *The Island of Dr. Moreau*. People have applied venom to monkey kidneys and lobster claws, to the iris of the eye of a rat and to the nerves of frogs and squid. They have poisoned rats, dissected them, liquefied heart, brain, spleen, liver, kidneys, lungs, and rump muscles separately, and injected them into other rats—all of which died except for those receiving the rump-muscle fluid. They have elicited venom from widows with electric shocks. They have given widows water laced with radioactive selenium and phosphorus and then counted the Geiger clicks in the organs of guinea pigs the widows killed. They have induced widows to bite laboratory rats on the penis, after which even the rats "appeared to become dejected and depressed." They have injected animals with the blood of human widow victims; the animals reacted as if they themselves had been bitten. In one experiment, scientists caused rats to be bitten on the ankle; then, at intervals, they amputated the bitten legs at the knee, to see how fast the venom spread. Only those who lost their legs in the first five minutes were spared the full effects of the toxin. Even those amputated in the first fifteen seconds showed some symptoms.

Such experiments revealed the peculiar reactions of different animals to the venom. Rats become more sensitive to noise, so that they're easily startled; they rub their snouts and twitch; they put their heads on the floor between their hind legs, as if expecting an air raid, before they die. Cats, those nocturnal hunters, come to fear the light. They crawl backwards, belly to the floor, howling, and then drop into a condition that in human schizophrenics is known as *waxy flexibility*. The animals remain catatonically still, holding any odd position the experimenter bends them into, before they, too, die. An early experimenter noted that cats exhibiting waxy flexibility don't react to being poked and cut. Among the animals who find widow venom especially deadly are guinea pigs, mice, horses, camels, snakes, frogs, insects, and spiders, including the widow itself. Others, like dogs, sheep, and rabbits, can often survive a bite.

The meager reactions of some animals left skeptics room for argument. The Russian government tried to resolve the question in 1899. Its experimenters couldn't provoke the spiders into biting, so they concluded the danger was mere folklore. The project's photographer apparently decided to illustrate this point by putting half a dozen widows on the naked chest of another man and taking pictures. During this stunt, the man being photographed got bitten. He was seriously sick within five minutes.

Meanwhile, at least half-a-dozen Western researchers

tried to toxify themselves. They teased widows into biting them, or else injected themselves with fluid derived
from the venom sacs of widows. All of these researchers
reported no symptoms at all—a result that bolstered the
position of the skeptics. Why weren't these men affected? Research in the decades that followed showed
an enormous variation in the widow's venom according
to environmental factors, especially season and temperature. The early experimenters may simply have collected spiders that were too cold or too old to produce
good venom. In the cases in which the experimenter
allowed himself to be bitten, rather than injecting an
extract, there's another possibility. The spider chooses
whether to inject venom, so she can deliver a dry bite if
she wants to. Doing so is sometimes a good strategy,
since the dry bite may succeed in driving off a big animal without any waste of venom. The men who injected themselves with extracts may have been misled
by some faulty chemical procedure.

In 1922 an arachnologist at the University of Arkansas, William Baerg, experimented on himself. At first he
couldn't convince the widow to bite him. Eventually he
did elicit a bite and was rewarded with three days of
pain and delirium in the hospital. That seems like com-
pelling evidence, but since other experimenters had
gone symptomless, the skeptics held out. In the next few
years the evidence mounted: a doctor compiled hundreds of case histories, and other experiments using re-

duced doses in the interest of safety produced slight symptoms.

The next researcher to risk the widow's bite was Allan Blair, an M.D. and a member of the faculty at the University of Alabama's medical school. Blair's wife and several others volunteered to serve as his guinea pigs in a widow bite experiment, but Blair declined their offers. Taking frequent measurements and thorough notes for the scientific article he would later write, Blair provided spectacular proof of the widow's power to harm human beings. His scientific triumph nearly killed him.

———

Look into the widow's face. This close, it doesn't even look black: it is glossed with light, supernally transformed into something luminous. A crown of black beads rims her head: she has eight eyes, though you cannot see them all as you stare into her face. The central pair looks blandly back at you. The exoskeleton is pitted and spined, nothing like the smooth, dark glass it appears to be from a distance. The forelegs seem to reach past you. Between them the hairy pedipalps dangle. Between the palps, the chelicerae: darker than the rest of the face, each shaped like the outline of a hacksaw, each terminating in a fine pale fang that looks like a cat claw and curves in toward the middle. The chelicerae seem outsized, a Rip Van Winkle beard on the relatively small face.

You could never see the widow this way without some mechanical help. The face I'm looking at is a photograph taken in 1933, an extreme close-up of "Spider #111.33," as she was designated for research purposes. In the lower right corner of the photo is a handwritten note from the photographer to Allan Blair: "Lest you forget."

Blair had been keeping widows in his laboratory for experiments on animals. (One of his experiments proved even the widow's eggs are toxic to mice.) He and his colleagues and assistants had collected the spiders from the wild; widows were plentiful around Tuscaloosa, Alabama. Blair captured Spider 111.33 in a rock pile near his own home on October 25, 1933. Like the other captive widows in Blair's laboratory, she was kept in a jar and provided with live insects. A water beetle became her last meal before the experiment. Then she went hungry for two weeks. Since earlier experimenters, like Baerg, had sometimes found it difficult to provoke a widow into biting, Blair wanted his spider hungry and irritable before he made any attempt to get bitten. (Incidentally, two weeks without food is a cakewalk for a widow. Other scientists working with a similar setup—many numbered widows in jars on shelves—once found that they had misplaced one widow at the back of a shelf for nine months. When they found her, she was still alive and eager to eat.)

On November 12, Spider 111.33 was, in Blair's words, "of moderate size, active and glossy black, with characteristic adult markings"—he means the red hourglass—"and appeared to be in excellent condition." Blair described himself as "aged 32, weighing 168 pounds . . . athletically inclined and in excellent health." A former college football player, Blair had just won the university's faculty tennis championship. He had monitored his body for a week and found his condition "normal." He had no particular sensitivity to mosquitoes or bees.

At ten forty-five in the morning, Blair used a small forceps to pick Spider 111.33 up by the abdomen and place her on his left hand. Without being prompted, she immediately bit him near the tip of his little finger, "twisting the cephalothorax from side to side as though to sink the claws of the chelicerae deeper into the flesh." The bite felt like a needle prick and a burn at the same time. Blair let the spider bite him for ten seconds, the burning growing more intense all the while. He removed the widow, putting it back into its jar unharmed.

A drop of "whitish fluid, slightly streaked with brown" beaded at the wound—venom laced with Blair's blood. The wound itself was so small that Blair couldn't see it even with a magnifying glass.

Blair's right hand was busy taking notes. Two minutes after the bite, he recorded a "bluish, pinpoint mark" where he had been bitten; the mark was surrounded by a disk of white skin. The finger was "burning." Soon the tip of the finger turned red, except for the pale area around the bite. The pain became "throbbing, lancinating."

Fifteen minutes after the bite, the pain had spread past the base of Blair's little finger. The side of his hand felt a bit numb. The area around the bite was sweating. The pain quickly traveled up his hand and arm, but it still was worst at the tip of his finger, which had swollen into a purple-red sausage.

At the twenty-two-minute mark, the vanguard of the pain had spread to Blair's chest, and the worst of it had progressed to his armpit, though the finger continued to throb. Noting the pain in the lymph node near his elbow, Blair deduced that the toxin had traveled through his lymphatic system.

Fifty minutes after the bite, Blair realized that the toxin was traveling in his blood. He felt "dull, drowsy, lethargic"; his blood pressure dropped; his pulse weakened; his breathing seemed deep. His white count began the steep climb it would continue throughout that day and night. His blood pressure and pulse continued to worsen.

Soon he felt flushed and had a headache and a pain in his upper belly. Malaise and pain in the neck muscles developed. Blair turned the note-taking duties over to his assistants. Shortly after noon, he noted that his legs felt "flushed, trembly" and his belly ached and was "tense." A rigid, pain-racked abdomen is a classic black widow symptom, as Blair knew from his study of other doctors' cases. He must have suspected he was about to experience pain much, much worse than he already felt. He asked to be taken to the hospital, which was three miles away. The ride took fifteen minutes, during which, as they say in politics, the situation deteriorated.

At half-past noon, Blair was at the hospital. His pulse was "weak and thready." His belly was rigid and racked with pain. His lower back ached. His chest hurt and felt "constricted." "Speech was difficult and jerky," he wrote later, adding in the detached tone obligatory for the medical journal in which he published his results, "respirations were rapid and labored, with a sharp brisk expiration accompanied by an audible grunt."

Blair's pains made it difficult for him to lie down for electrocardiograms—in fact, an assistant dutifully wrote down that he described it as "torture"—but he managed to lie still, and the EKGs proved normal. Hearing about the painful EKGs later, newspaper reporters wrongly assumed the venom had injured Blair's heart. That myth was repeated and embroidered in the press for decades, giving the widow's danger a spurious explanation easier for casual readers to grasp: heart attack.

Two hours after the bite, Blair lay on his side in fetal

position. The pain had reached his legs. His "respirations were labored, with a gasping inspiration and a
sharp, jerky expiration accompanied by an uncontrol-
lable, loud, groaning grunt." He could not straighten
his body, which was rigid and trembling; he certainly
couldn't stand. His skin was pale and "ashy" and slick
with clammy sweat. In short, he had fallen into deep
shock. The bitten finger had turned blue.

Folk remedies reported from places as diverse as
Madagascar and southern Europe involved the use of
heat, and some doctors had reported hot baths and hot
compresses helpful. William Baerg had attested the
pain-relieving power of hot baths during his stay in the
hospital. Blair decided to try this treatment on himself.
As soon as his body was immersed, he felt an almost
miraculous reduction of his pain, though it was still
severe. His breath laboring, his forearms and hands
jerking spastically, he allowed a nurse to take his blood
pressure and pulse. His systolic pressure was 75; the
diastolic pressure was too faint to determine with a cuff
and stethoscope. His pulse remained weak and rapid—
too rapid to count.

Forty-five minutes after Blair had arrived at the hospital, his colleague J. M. Forney arrived to take care of
him. Forney found Blair lying in the bathtub, gasping
for breath, his face contorted into the sweat-slick,
heavy-lidded mask that has since come to be recognized
as a typical symptom of widow bite. Blair said he felt

dizzy. Forney later commented, "I do not recall having seen more abject pain manifested in any other medical or surgical condition."

After soaking for more than half an hour, Blair was removed from the bath, red as a boiled lobster. His breathing, like his pains, had improved as a result of the bath. Fifteen minutes later, both the ragged breathing and the pain were back at full force. Blair writhed in the hospital bed. Hot water bottles were packed against his back and belly, again reducing his pain. Perspiration poured from him, drenching his sheets. His blood pressure was 80 over 50. His pulse was a weak 120. He accepted an injection of morphine to help with the pain.

Blair continued to gulp down water. Sweat poured out of him and would for days, leaving him little moisture for producing urine. A red streak appeared on his left hand. He vomited and had diarrhea; he couldn't eat. In the evening of the first day, his blood pressure rebounded to 154 over 92; it stayed high for a week. His face swelled; his eyes were bloodshot and watery.

The night was terrible. He felt restless and could not sleep. The pain persisted. He had chills. A dose of barbiturates didn't help. He was in and out of hot baths all night. Sometime in that night the worst part came. Blair felt he couldn't endure any more pain. He said he was about to go insane; he was holding on only by an effort of his steadily weakening will. His caregivers injected him with morphine again.

The next day, his hands trembling, his arm broken out in a knobby rash, his breath stinking, his features distorted by swelling, Blair was still in pain, but he knew he was getting better. In the evening, as he sat guzzling orange juice, sweat pouring from his body, his worst symptom was pain in the legs.

By the third day, Blair was able to sleep and eat a little. His boardlike abdomen had finally relaxed. He was beginning to look like himself again as his swollen face returned to its normal proportions. He went home that day. It took about a week for all the serious symptoms to vanish. After that, his body itched for two more weeks, and the skin on his hands and feet peeled as if burned.

Blair later returned to his native Saskatchewan, where he had an illustrious career in cancer treatment and research. When he died of heart trouble at age forty-seven, prime ministers and other public figures eulogized him. The story of his black widow experiment, which the wire service had named one of the top ten human interest stories of 1933, was retold in the papers at his death, and one more accretion of myth was added to the story when his heart trouble was falsely attributed to the bite of the black widow sixteen years before.

Blair's ordeal convinced the skeptics the widow's bite is toxic and potentially deadly. Thousands of cases of latrodectism, as widow poisoning is called, have been documented since then. The variation in symptoms from one person to the next is remarkable, making some cases hard to diagnose. The constant is pain, usually all over the body but concentrated in the belly, legs, and lower back. Often the soles of the feet hurt—one woman said she felt as if someone were ripping off her toenails or taking an iron to her feet.

Some doctors trying to diagnose an uncertain case ask, "Is this the worst pain you've ever felt?" A "yes" suggests a diagnosis of black widow bite. Several doctors have made remarks similar to Forney's, about the widow causing the worst human suffering they ever witnessed (though one ranked the widow's bite second to tetanus, which is sometimes a complication of widow bite). One of the questions Blair had in mind when he began his experiment was whether people acquire immunity over successive bites. He never answered this question because, as he frankly admitted, he was afraid of having another experience like his first.

Besides pain, several other symptoms appear regularly in widow victims, and Blair's suffering provided examples of most of them: a rigid abdomen, the "mask of latrodectism" (a distorted face caused by pain and involuntary contraction of muscles), intense sweating (the body's attempt to purge the toxin), nausea, vomiting, swelling. A multitude of other symptoms have occurred in widow bite cases, including convulsions, fainting, paralysis, and amnesia. Baerg and a number of

other victims reported nightmares and sleep distur-
bances after the life-threatening phase of their reactions
had passed.

Blair's fear for his sanity was not unusual either.
Other patients have expressed similar fears, and some,
like Baerg, have lapsed into delirium. Some have tried
to kill themselves to stop the pain. (A few people have
intentionally tried to get bitten as a method of suicide. It
would be hard to imagine a method at once so uncertain and so painful.)

The venom contains a neurotoxin that accounts for
the pain and the system-wide effects like roller-coaster
blood pressure. But this chemical explanation only
opens the door to deeper mysteries. A dose of the
venom contains only a few molecules of the neurotoxin,
which has a high molecular weight—in fact, the molecules are large enough to be seen under an ordinary
microscope. How do these few molecules manage to
affect the entire body of an animal weighing hundreds
or even thousands of pounds? No one has explained the
specific mechanism. It seems to involve a neural cas-
cade, a series of reactions initiated by the toxin, but with
the toxin not directly involved in any but the first steps
of the process. The toxin somehow flips a switch that
activates a self-torture mechanism.

People sometimes die from widow bites. Thorpe and
Woodson report the case of a two-year-old boy who was
walking in the garden with his grandfather when he

said his big toe hurt. He soon fell unconscious. Within an hour he lay dead. The grandfather went to the spot in the garden where the boy had felt the pain. He turned over a rock. A black widow, suddenly exposed, wobbled away over the flagstones.

Widow bites kill old people with greater-than-average frequency, apparently because they're especially susceptible to some of the secondary effects. The high blood pressure, for example, kills some victims via stroke or heart attack. That's what happened to Harry Carey, an actor best known for his character roles in John Wayne Westerns. A black widow bit him while he was working on *Red River;* he died of a heart attack.

Many of the symptoms reported for widow bites are actually symptoms of such complications. Anybody who already has a serious medical problem runs a big risk when bitten by a widow. One man with a chronic kidney problem died from a bite, the toxin overtaxing his diseased kidneys as they tried to clean his blood. Another common complication, and a proven killer in widow bite cases, is infection. The widow's habit of dwelling in outhouses and piles of trash can make her bite septic. Besides tetanus, encephalitis and gruesome staph infections of the skin have also killed bite victims.

———

Some early researchers hypothesized that the virulence of the venom was necessary for killing scarab beetles.

The scarab family contains thousands of species, including the June beetle and the famous dung beetle the Egyptians thought immortal. All the scarabs have thick, strong bodies and tough exoskeletons, and many of them are common prey for the widow. The tough hide was supposed to require a particularly nasty venom. I have seen widows take dozens of thumb-thick American-style dung beetles. These broad-shouldered creatures, smaller but still massive replicas of their African cousins, are armed with digging claws like ornate hair combs on their front legs. They come wobbling along the gutters and sidewalks around my house in the summer. I remember hitching a toy wagon to one when I was a child: he was equal to towing a load several times his weight. When I tired of him and threw him into a widow's web, he struggled in his windup toy way and threw a defensive flurry of tarry feces. The patient widow hung just out of range and threw silk onto him for more than an hour before moving to his front end to deliver the killing bite. Big as the dung beetle was, her bite, once delivered, killed him in minutes.

As it turns out, the widow's venom is thousands of times more virulent than necessary for killing scarabs. The whole idea is full of the widow's glamour: an emblem of eternal life killed by a creature whose most distinctive blood-colored markings people invariably describe as an hourglass.

No one has ever offered a sufficient explanation for the dangerous venom. It provides no clear evolutionary advantage: all of the widow's prey items would find lesser toxins fatal, and there is no unambiguous benefit in killing or harming larger animals. A widow that bites a human being or other large animal is likely to be killed. Evolution does sometimes produce such flowers of natural evil—traits that are neither functional nor vestigial, but utterly pointless. Natural selection favors the inheritance of useful characteristics that arise from random mutation and tends to extinguish disadvantageous traits. All other characteristics, the ones that neither help nor hinder survival, are preserved or extinguished at random as mutation links them with useful or harmful traits. Many people—even many scientists—assume that every animal is elegantly engineered for its ecological niche, that every bit of an animal's anatomy and behavior has a functional explanation. However, nothing in evolutionary theory sanctions this assumption. Close observation of the lives around us rules out any view so systematic.

We want the world to be an ordered room, but in a corner of that room there hangs an untidy web. Here the analytical mind finds an irreducible mystery, a motiveless evil in nature; and the scientist's vision of evil comes to match the vision of a God-fearing country woman with a ten-foot pole. No idea of the cosmos as elegant design accounts for the widow. No idea of a benevolent God can be completely comfortable in a widow's world. She hangs in her web, that marvel of design, and defies teleology.

MANTID

From my second-floor apartment I could see across the parking lot to the creek, and I used to step out on the landing at dusk to watch the fireflies lighting up against the backdrop of the darkening pines and maples and Osage orange trees. One night the miller moths were especially thick around the light fixture on the landing, and I was about to go inside because the furry, knuckle-sized creatures kept bumping me, leaving iridescent streaks of dusty scales on my sweating skin. That's when I noticed one of them jerk from the arc of its flight and buzz like a disgruntled bee.

There was a beige-painted wood banister along the landing, and a piece of it had grabbed the moth and was chewing its head off. As I looked closer, the carnivorous piece of banister adjusted its grip slightly, and I recognized it as a praying mantis, or mantid, as the entomologists prefer. She held the moth, wings down, before her face and turned to stare at me. She looked like a person wiping her face with a napkin.

The mantid was two and a half inches long and exactly the color of the banister. Her triangular head came to a point in mandibles like two tiny pairs of pruning shears; they were surrounded by four fingerlike palps. She walked on four legs. The front pair of legs, the ones she didn't use for walking, were covered with spikes and ended in boat hooks, and she held these up before herself. The odd position of mantid forelegs has suggested contemplation or wisdom to many people in different parts of the world. The Greek root of *mantid* means "prophet." In Africa and the Middle East, legends of religious and prophetic significance adhere to the mantid. In the United States, its common names include *soothsayer*. And, of course, it is accused of "praying."

I trapped the mantid in a gallon pickle jar and brought her inside, adding a few twigs and leaves for her to climb on. By morning she had turned green to match the leaves.

I caught a few miller moths and tossed them in. The mantid climbed halfway up a thick twig and clung there with her middle and hind legs, her big forelegs folded close to her chest. A moth flew near. Her head swiveled to watch its erratic, glass-bumping flight.

She snatched it from the air. I didn't see it happen; her strike was too fast to see, even as a blur. Scientists say an entire strike lasts one twentieth of a second. I only sensed some startling occurrence, and then the moth was trapped in her spiky arms. She was already biting it in the furry scales just behind its bald head. Mantids generally bite in just this spot, severing the prey's major nerve, the equivalent of a spinal cord. This surgical technique, which mantids somehow instinctively apply to a wide range of prey, breaks the connection between an insect's limbs and brain. It's not necessarily a fatal wound, but it leaves the insect powerless to defend itself.

The moth flapped its wings into a buzzing blur every few seconds while the mantid unhurriedly ate it, starting from the head. The pruning-shear mouthparts worked away, biting out chunks of moth and lapping the juices. The moth's scales, which had broken into particles of dust when they smeared my hand, looked like little brown feathers when they were whole, and they drifted down in a steady snow.

I kept the mantid for a week or so, frequently feeding it moths. The twig it perched on was unsteady. Sometimes it spun out of place when the mantid struck at a moth. The mantid's strike, missing its target as the mantid lost her footing, would hang in the air for an instant, giving me a rare look at the process—the arms unfurled, reaching, like a model showing evening gloves.

I had read that mantids eat almost anything, from hornets (they leave the stinger uneaten) to humming-birds to frogs. One mantid seized a mouse and ate it alive, starting from the nose. (That mantid was five inches long.) I myself had seen them eating black widow spiders; as the mantid devours what passes for the spider's brain, one spider leg moves up and down as if keeping time to music. I fed this mantid whatever crawled across the landing: a spotted white caterpillar, which held to the grass stem from which the mantid plucked it, bending the stem almost double on its way to death. House flies—she ate only the larger ones. Field crickets that walked up to her boldly like paunchy men in tuxedos. A huge orb-weaving spider with legs striped in silver and gold. She would eat anything she could see moving; I watched her watch the movement before she made the kill.

The mantid is a visual animal, far more so than al-most any other arthropod. Her two huge eyes form a human-style face: gaze at her and she seems to be gaz-ing back, as cats and monkeys do. Try the same trick with most insects and, if you can even discover any eyes, you'll find they don't give the same impression.

The mantid's two big eyes, arranged so that both can see forward, give her stereoscopic vision. That means she can see two images of the same thing and, by combining the two, judge depth. That's the same trick we humans use. The mantid can also see a little bit of color. But her specialty is seeing motion: in order to eat, she has to detect animal motion among wind-stirred leaves. When she sees prey moving, she freezes until it comes close; then she launches her invisibly fast strike. She can see in light or near-dark, but, like many predominantly visual animals, she prefers to hunt by day.

The mantid has a feature unique among insects: the ability to turn her head. A mantid can actually look over her own "shoulder." This combination of traits— swiveling head, stereoscopic vision, depth perception, and motion detection—causes mantid behavior to resemble that of cats and people more than it does typical insect behavior, at least in matters of food and self-defense. For example, what happens if you thrust your hand toward a person's face, stopping just short of contact? The first reaction is a flinch. What happens if you try the same thing with a cockroach, a close cousin of the mantid? It runs, changing directions frequently to confuse you. And what if you try the same thing with a mantid? She flinches. Mantids react like people because they see the world in basically the same way.

Visual talents of this sort usually go with a predatory lifestyle. We're not pure predators like the mantid, but we have the equipment to be. We also have grasping appendages, another frequent predator trait, just as cats have grasping claws and mantids have what scientists call *raptorial* arms. "Raptorial" is biologese for "grasping." Few insects can hold prey items as the mantid can.

All of our similarities to the mantid result from convergent evolution, which means that unrelated animals develop similar features because they've adapted to similar environmental challenges. The killer whale, a warm-blooded animal built on a frame of bones, is shaped like the white shark, a cold-blooded animal with a cartilaginous skeleton. That's because they both live in the ocean and eat seals: convergent evolution. Our similarities to the mantid are subtler.

To start with, we both have weak senses except for sight. If you want an illustration of the weakness of your ears and nose, follow a dog around outdoors and try to figure out what he's alerting to every time he cocks his head or stops to stare. You'll soon believe yourself deaf and blunt-nosed. If you tried the same thing with a mantid, you would understand her better, even though the dog is, relatively speaking, your first cousin and the mantid a stranger.

The metaphors reveal our visual nature. When we English-speaking humans want to show that we understand something, we say "I see." But "I hear" is the language of rumor; it means something is possible but unproved. I'm told many languages have analogous sensory metaphors. They reflect the epistemology our bodies teach us.

The mantid is pretty useless when it comes to hearing. Of the approximately two thousand mantid species in the world, many lack ears altogether. In only a few species are both genders equipped with ears. The remaining species—over half the total—exhibit sexual dimorphism. The male has an ear, and the female doesn't. The mantid ear is unique; each male has only one, which is in the center of his chest. Most animals can tell the direction of a sound because of subtle differences in reception between their two ears, but the mantid lacks this talent. He cannot use his ear to locate food or mates. His vision is, of course, all he needs to hunt prey. He finds receptive females by scent: the females emit pheremones.

To follow this scent, the male flies by night. This fact accounts for his longer wings. The females of most mantid species don't fly at all. This is where the male mantid's ear reveals its function. It detects only high frequencies, so it is useless for most defensive purposes, but it does pick up the echo-locating screams of bats. Insectivorous bats eat mantids on the wing. A mantid that hears a bat's call power-dives to avoid being taken. Since this tactic doesn't require the mantid to know where the bat is, the single ear suffices. Scientists believe the mantid's ear, for which no other function has yet been discovered, evolved in response to predation by bats.

The mantid as meal: that brings us to another important trait we share. We're both predatory animals in the middle of the food chain. The mantid is built to kill, but she can also be killed, and often is.

———

We sat talking on the porch, and out in the grass at the edge of the light the black cat continually pulled himself into a tight ball and then sprang at some insect floating by. We saw him miss a few moths, and we saw him leap at things we couldn't see. After a while he came trotting up to the porch with something in his mouth. The something was thrashing its legs in the fur of the cat's cheeks.

The cat crouched to play with his captive, dropping it on the sidewalk and pinning it with one paw. It was a big green mantid. When the cat raised the paw and looked, the mantid rose on his hind legs, throwing his formidable front limbs into the air to show their red and yellow undersides, and staggered toward the cat, as if to intimidate the feline with its size. Scientists call that a *threat display*. The cat clapped his paws together on the mantid.

A second later the mantid slipped free and burst into buzzing flight, making a swift, clumsy arc before the cat's face. The cat sprang to catch the mantid in midair. And when he had his captive wrestled to the cement, he was through playing. He bit and pulled his head back, breaking the mantid in half. The fight went on for another five minutes or so, the black cat eating, the green mantid still waving his limbs in protest. The cat left the spiny forelimbs and a tangle of winged thorax.

————

The human animal, too, is in the middle of a predatory chain. This idea ticks off a lot of people who, generally because of some religious or cultural bias, think we ought to be the bosses of the animal kingdom. ("Kingdom": even that bit of taxonomic apparatus shows a human bias toward thinking in hierarchies.)

A skeptical reader may point out that, as he sits in the comfort of his den in the middle of his town reading this book, he's in virtually no danger of being eaten by anybody. Good for you, sir; you rank with the termite. We humans are safe in our own shelters and towns, just as termites are safe in their mounds.

Another reader remarks that her shotgun elevates her to the status of top predator. You're right, ma'am; our talent with projectile weapons is a powerful one, and puts us on a par with the rock-chunking baboons. If, however, you encountered any of the predators that really see us as food—not your flimsy North American predators like the coyote or the cougar (though each of those has killed a human or two), but, say, a saltwater crocodile or a white shark—your gun would stand only a slim chance of winning the day for you.

But, a third reader murmurs sullenly, we can kill anything if we team up on it. True, and we often use this trick to eradicate troublesome tigers. They become troublesome by eating a few people—in some cases, a few hundred people. Similarly, baboons pack-hunt leopards, but this doesn't change the fact that leopards are serious baboon-eaters.

The leopard is adept at shaking the anthropocentrist in all of us, at making us rethink our preconceptions of hierarchies in the natural world and especially our sense of sitting at the top of the heap. The fossil skull of an early man shows the puncture wound characteristic of death by leopard. This cat is far smaller than the lion or the tiger, but he kills far more people. He's been with us, in Asia and Africa, since before we wrote histories; he was there before we came out of the jungle to walk the savannas, and he's never shared our conceit that we're fundamentally different from other apes and monkeys. Today he eats our close relatives, chimpanzees and baboons, just as he ate our ancestors. He still eats us—hundreds of us every year. Some single leopards have killed hundreds of humans. The Panar leopard of northern India, for example, took more than four hundred people. That's the Indian government's official death toll, which strictly excludes uncertain cases. Leopards share our enjoyment of sport killing. One in Tanzania killed twenty-six people without eating any of them.

There's a theory that lions and tigers eat people only after they learn to like the taste of human flesh by scavenging battlefields and plague-ridden villages. Whether that's true of lions and tigers or not, it certainly doesn't apply to the leopard, which needs no prompting to add some protein-rich primate to its diet. It's also claimed that the big cats resort to human-eating when they are too old and feeble to catch their "natural" prey. But among human-eating leopards, healthy, exceptionally large males are common.

The lion and tiger kill by seizing the throat or nose and closing down the flow of air with a patient vise of a bite. This strategy is well suited to hoofed animals, which, once brought down, have little defense. The leopard rarely suffocates his prey. He is used to hunting animals that can scratch and grab at his eyes—baboons, chimps, humans. He has no patience for a caught animal. He bites through the skull for a quick kill.

Like the mantid's, his strategy aims to break the nervous system. He is built to kill us.

————

Our mythic prototype of a creature from another planet has round luminous eyes on a face whose other features seem atrophied. That's the face of a mantid.

Things that are really alien don't look alien. Look at diatoms under a microscope and you don't have the feeling of looking at once-living things; the creatures more nearly resemble wallpaper. Do you empathize with a trout, a sparrow, or a lobster? Forms of life that are relatively different from us cause us little or no visceral response. Animals that resemble us more closely—cats and dogs, for example—make some of us feel sympathy, friendship, even love. We understand some of these animals' motives, because we share them—the visceral pleasure of eating, games based on instinctive hunting and fighting habits, and so on.

The things that scare or repulse us are those that are sympathetically human in some respects, but markedly alien in others. For example, apes appear in a disproportionately large number of horror fantasies, from "The Murders in the Rue Morgue" to *King Kong.* Apes disturb us with their imperfect humanity. So do dolls. In talking with people about their childhood fears, I have heard many mentions of dolls and mannequins, and the fear seems to center on the eyes: the rest is passably human, but the dead eyes make the thing terrible, at least in the dark. Of course, the human body itself becomes an object of terror as soon as it dies, because it is still human, but not in the way we want it to be.

That archetypal image of the extraterrestrial—bilaterally symmetrical, bipedal, visual—is an unlikely choice if you want to imagine what might really live on other planets. After all, what are the chances of an alien looking almost exactly like *any* earth animal, and us in particular? But the image makes sense psychologically, for it is us touched up with a few strokes of strangeness.

The mantid spends most of her time looking like a reasonable enough, if slightly sinister, character. She walks along poised to grab something; she reacts visibly to prey; she will try to avoid anyone who offers to step on her. If you persist in threatening her, she will throw her arms in the air and wave them, as if to say, "See how big I am? You'd better think twice about dogging me, buddy." But sometimes the mantid will show you that, however understandable most of her actions may be, she also has a side alien to us.

———

The female is the color of jade, her abdomen thick and fleshy. She is mostly still, her feet hooked into the texture of the elm tree's bark. When I first glimpsed her, I mistook her abdomen and wings for the chrysalis of a monarch butterfly. But as my eyes picked her shape out of the relief map of bark and beetle-chewed leaves, I saw the long stretch of her thorax ending in a plow-share head, at the upper corners of which her amber eyes bulge like beaded water, each of them punctuated with a black period in the middle that resembles a pupil but is not.

The male approaches, angling away from her face to avoid being too easy a target. He is more slender, and holds his hind wings slightly ruffled beneath his green forewings; his angular looks could help him pass for a twig. His eyes are different too, duller, more ascetic. He lacks her mass but is almost as long, perhaps two inches. He walks down the trunk toward her, his body held away from the bark by his four hind legs, which jut out to the sides of his body from his thorax, then turn right angles to meet the bark. His head swivels as he comes, keeping her in view. She watches him, and her mouthparts work idly.

When he is perhaps an inch away, he stops and begins to sway on his bent legs like a sumo wrestler in warm-up. Her forelegs unfurl slightly, then stop in midair, one slightly ahead of the other. She, too, begins to sway. He walks from side to side before her, sometimes stopping to sway, his wings unfolding slightly and trembling. She watches him walk. Her own movements stop, or perhaps continue too subtly for me to see. He edges around to her side.

Suddenly he runs a good six inches and lunges into the air, his blunt forewings flicking forward to let the transparent hind wings fan out into buzzing flight. She turns her head to follow him with her full-yellow-moon eyes. He lands in the grass, then flaps back to repeat his flying leap. This time he returns and begins to slap the female with his antennae. His head moves from side to side like that of a playful dog fighting his master for a toy. The slender antennae lash her like the proverbial wet noodle, failing to even ruffle her antennae.

She strikes.

Now she is standing still, her blur of motion over so

quickly it might seem unreal, except that she is slowly eating the right half of his head.

He stands swaying, his actions only slightly interrupted by the amputation of half his head.

Then, while she is still eating, he crawls onto her back. He seems in this semiheadless state to have found a renewed vigor and sense of purpose. There will be no more showy stunts. His pale penis emerges from the rear of his body, extruded between the plates of his exoskeleton. His abdomen snakes around beside hers and forms a painful-looking curve. They begin to copulate.

Turning her face almost 180 degrees, she regards him for a moment, as if his attentions were a distasteful surprise. Then, twisting with some difficulty, she brings her raptorial forelimbs into position and strikes again. This time she retrieves the remainder of his head and a scrap of his thorax, from which one foreleg dangles.

He doesn't seem to mind. He stays on her back like some undersized Headless Horseman. I recall my grandfather once showing me a big female mantid and calling it a "devil-horse."

The copulation continues. It lacks the aerobics of a mammalian encounter. After the insertion it involves, besides the cannibalism, merely clinging and a slight pulsing in the male's soft abdomen. It may go on for a long time; some couplings have outlasted my patience for watching. The genitals fit so tightly that, if you try to separate the pair, their bodies will tear apart before they disengage.

The female hasn't finished her meal. She strikes again, removing everything forward of his middle pair of legs. She eats rapidly. His raptorial forelimbs lie on the bark like discarded hand tools. She walks out from under what's left of his body and stands a few inches away, cleaning her forelimbs with her mouth. The male's remains crabwalk a few steps. The abdomen pulses faintly. The female picks her steps on the rough bark as she goes away. He stays there, wiggling his abdomen obscenely, staggering in sideways arcs. He will do so until something else comes along to eat him.

Males die a few days after copulation, even if the female hasn't harmed them. The female will lay eggs in a day or two. She lays them in a gluey substance she squeezes out of her abdomen, all the while moving herself in a spiral like a cake decorator's bag of icing. Special appendages at her rear end whip the substance into a froth. One egg case takes her a whole morning, and by afternoon the gluey stuff has set like cement. She usually gets at least three cases built before she dies, each containing up to two hundred eggs. Each one looks like a little army barracks. A case is impervious to just about everything except the teeth of rodents and the mandibles of parasitic wasps.

The hardwiring for the entire mating ritual lies in a cluster of nerves in the floor of the thorax. The brain is

not involved, except to inhibit the mantid from constantly going through the mating motions. That's why
the male not only can finish without a head, but even
performs with more gusto once he's decapitated. The
female can mate headless as well, though that's rarely
necessary. The female can even lay her eggs after she
loses her head. The cockroach, a cousin to the mantid,
has the same peculiar wiring. It has long been known
that roaches are capable of learning; they can run mazes
and can even be conditioned to flee darkness and love
light. This latter exercise has been replicated with headless cockroaches. They first learn the experiment after
their heads have already been removed, and they repeatedly show that the learning has taken. Their learning ability is not in the head.

Perhaps you wonder how the roach can survive without a head. Well, it does need its head for eating. After
a few weeks, a headless roach starves to death.

The mantid, which depends on her eyes and specializes in severing a prey animal's brain from the rest of its
nervous system, can survive the devastation of her nervous system and the amputation of her eyes.

Alien, indeed, from the human perspective. Yet, some
of the control mechanisms for human lovemaking are
low in the spine, not in the brain.

———

In France, folklore has the mantid pointing lost children toward home. Zulu legend depicts the mantid as a stealer of children. Those two opposite characterizations show how readily we attach motives to the creature whose moves so often resemble our own.

Of course, those two legends also show how silly our anthropomorphic explanations can be. Early entomologists often described the mantid as a hypocrite because she acts as if she's praying while she's really plotting the murder of some hapless bug. This is exactly the sort of foolishness contemporary scientists hope to forestall when they advise us not to anthropomorphize at all. Other animals may or may not have mental processes like thought and emotion, the biologists say; it's best not to assume, and of course we can't observe such phenomena directly. This position too often gets oversimplified, so that a lot of people recite the "fact" that animals have nothing like human emotion. This idea, common as it is among educated people, is a misreading of the attempt at scientific objectivity, which merely asks that we suspend judgment on the question until we have proof.

We may not have proof, but we do have good evidence of emotional lives in mammals. Apes trained to use sign language sometimes overflow with emotion, saying things along the lines of "You are an unpleasant excrement-head." Of course, any lover of cats or dogs takes an emotional empathy with these mammals for granted. The theory among some scientists these days is that the emotions of unity—love and the like—developed with the mammal brain, and that the more primitive reptile brain is limited to aggression and other simple feelings.

But what about animals even more alien to us—more "primitive," as our egos would have it—than reptiles? Do they feel? Science tends to treat these creatures as electrochemical machines. I have had a hint or two of something deeper in my dealings with arthropods. One particular incident made me wonder all over again where the lines are drawn. It also reminded me of a certain overworked quotation—Hamlet to Horatio, about the things in heaven and earth.

———

It was the rain that drove them up into the daylight world.

In the semiarid region where I live, these beasts must be plentiful underground, but I rarely see them, even when I'm looking under rocks and boards for interesting creatures. The beast's shape marks it a relative of the cricket, but its back is humped, its forelegs are thicker. There are many such creatures—camel crickets, mole crickets, Jerusalem crickets, all burrowing, seldom-seen inhabitants of the soil. The Jerusalem cricket has a bulbous, disturbingly humanoid head that accounts for the common name *child-of-the-earth*. But the beast I'm talking about doesn't completely match the looks and behavior of any of these well-documented insects. Doubtless some entomologist has cataloged it, but I have never found it in a book.

The beast is a gleaming red-brown—it looks as if it might be made of the kind of plastic used for tortoise-shell combs and brushes. I had seen them at the bottom of water meter wells three or four feet underground, and I had seen small specimens above ground a few times, always in wet weather. But I had never seen anything like this.

It was late summer in the wettest year I could recall. Much of the country was under flood. We weren't flooded in our semidesert, but the black earth had grown a leprous infection of white mushrooms, and every outdoor thing seemed transformed by the wetness into a refractor of rainbows. The world stank with rot and rebirth—a smell delirious and nauseous.

As I pulled my car into the wet driveway one afternoon, I saw the beast crossing the cement. While I was still in the car twenty feet away, I recognized it as the species I'd seen a few times before, but I hardly believed it. The thing was larger than some adult mice I've seen. It was the third one I had seen during that wet spell, but the other two had been much smaller. It moved slowly, walking like some deliberate beetle, not jumping as a true cricket would. The proximity of my car disturbed it not at all. I had been hunting rattlesnakes and had several jars in the car for collecting the snakes' heads and tails. I used one of these to catch the cricket-beast, which walked agreeably into the jar with no urging from me.

I had no special plan for the cricket-beast. I didn't even know what it was. After checking a reference book, I tentatively decided it must be a type of camel cricket that eats rotting vegetation.

I transferred the thing to a gallon jar half-filled with dirt. I threw in some wet leaves for food. It moved its slow body, heavy as a brass bullet, through the leaves. My family came to the consensus that it was one of the most disgusting things I had ever brought in. Having nothing better to do with it, I figured I would keep it around to look at for a day or two, then feed it to one of the tarantulas that occupied terraria on my utility room counter.

That evening, the rain was at it again, and a mantid squeezed in the back door. He was about two inches long, and gray. I decided to see whether this thin, perfect predator could handle something as large as the cricket-beast. I dropped the mantid into the gallon jar.

The mantid did something I had never seen before. He looked at the cricket-beast and began to run away sideways, keeping his face toward the beast at all times. He seemed scared. Of course, even as I thought this, I accused myself of anthropomorphizing. I still thought the cricket-beast was an eater of plant rot. I recalled dangerous situations in which I had seen mantids—tangled in a black widow's web, battered by a cat, swarmed by ants, shoved into a jar by yours truly. In none of these situations had I seen a mantid use this body language of apparent fear.

I made chicken noises at the mantid.

The cricket-beast, waving its long antennae, turned toward the mantid. The mantid froze. The beast held its position in the center of the round jar. This is the top-predator position—a tarantula placed in a similar environment will, after a little exploration, take up this same spot in the middle of a container. I said as much to my wife, her sister, and her sister's husband, explaining that the cricket-beast must be an idiot not to realize what danger it was in.

The mantid took another step away. The cricket-beast, which had been sluggish up to now, leaped. The mantid was knocked wings over teakettle, landing a few inches away. As he tried to regain his "feet," the beast pounced again. This time it landed squarely on the mantid and bear-hugged him. Then it began to eat the mantid in a leisurely but methodical way, its many mouth parts wiggling like fingers. It chewed the mantid's face off first, and continued downward, not even pausing at the thick carapace. The four of us watching were amazed and repulsed. The others were not avid bug-watchers, as I am, but the spectacle was so intense in its microcosmic way that no one could stop looking.

In ten minutes the mantid was gone. Nothing remained but his transparent wings. The cricket-beast crawled sluggishly to the center of the jar.

RATTLESNAKE

It lies half-coiled in a stand of dusty green weeds, its jaw against the ground to catch the vibrations of any moving thing. Its body, patterned with the colors of dead grass and earth, is touching a stack of iron pipe. Its forked black tongue slips out of its closed mouth, slashes in several directions, and slips back in (think of a person sucking down a strand of fettuccine). It is licking up particles of airborne scent and brushing them against the mass of olfactory nerves in the roof of its mouth. Its pupils, which would be only slits in the sun, have ballooned in the near-dark.

The rattlesnake has followed a scent along the pipes, and here it stops to wait at the turning of a scent-path. The prey, whatever it is, has the habit of following the shape of these discarded pipes. Probably it has a nest among them. The rattlesnake is still except for its active tongue, which slides out every few seconds, invisible in the dusk except for its gleam.

I don't see the field mouse arrive. He is suddenly there, tentative in his movements, a run of a few inches and then a pause. His coat is pale brown on top and white on bottom; his eyes are the sleek brown of apple seeds. He stops, runs to one side, stops, runs back the other direction, stops and rises on his hind legs. He seems to know something is wrong, but probably that's my imagining. I look to the snake and can't see it—only dirt and weeds and scraps of iron. I blink a few times and there it is in exactly the same place, my eyes and brain finally interpreting its pattern.

The mouse is a few inches from the safety of the pipes, but he darts around in the open. Does he smell the snake? I can't decide whether I smell it or not. The mouse runs onto a higher clump of dirt to look around and sniff. But it's not a clump of dirt. He is standing on a thick loop of the snake. The snake does not move.

The mouse comes down and moves away from the pipes. A blurred movement, the rustle of one weed— something happens too fast for me to see. The mouse leaps into the air but makes no sound. He lands on his feet, takes a step or two, defecates, and stands shivering. The snake slides back a few inches. It isn't moving now, but it's watching with the heat-sensing pits below its eyes. Its strike, gauged by means of the pits, has hit home. The pits work with heat as human eyes do with light, creating stereoscopic "vision" and thus a fine discrimination of direction and distance.

The mouse rolls on his side, breathing heavily, spasms rocking the forelegs and head. The snake waits. After a long while it slips closer. Its tongue runs over the mouse, which is still twitching. The snake makes a half-circle and settles its head near the mouse's. The mouse still appears alive to me, but the snake has its heat-sense and may know something I don't. Alive or dead, the mouse has already begun to be digested. The venom is breaking down cell walls; tissues are flooding with blood; the flesh is softening.

The rattlesnake swallows the mouse head-first, the hollow tube of its glottis pushing to the front of its mouth at the bottom so that it can breathe while it eats, its delicate bones momentarily separating, its muscles working and rippling. The swallowing is a long process; the mouse remains partly visible for perhaps five minutes. Before he disappears entirely, I see a hind leg twitch, and then for a long while only the mouse's dark tail hangs out, and then it is gone.

———

Two men were hunting in the woods in midwinter. They came upon a clearing where a seven-foot Eastern diamondback lay soaking in the sun, sluggish in the cold. The hunters knew of a man in town who made money by milking the venom from rattlesnakes; he sold the venom to a pharmaceutical company for use in making antivenin. They decided to catch the snake and sell it to the venom collector.

Catching it was easy. The snake was cold and slow. One man grasped it behind the head and around the middle. He knew it couldn't bite him in this position. They got into the pickup and headed for town.

The heater was on in the pickup.

They reached town and were about to get out of the truck. The man holding the snake had grown complacent on the ride to town; the snake had grown warm. The man must have relaxed his grip for a second. The snake whipped suddenly, too fast to react to. Its two fangs punched into his arm.

The swelling started minutes after the bite, while they were driving to the hospital. Eventually his arm purpled and gleamed with the sheen of leaking plasma. It grew to Popeye proportions; they had to cut the shirt to get it off. The skin ripped open. He bled from his mouth and nose and from the pores of his skin. His arm broke out in tiny blisters. He was in intensive care for four days, a nightmare time in which the doctors used calm, soothing tones to discuss his "hypotensive crisis" and the amputation of his arm. Finally, in a sort of Faustian bargain, he was allowed to keep the arm in return for the sacrifice of his hand. He handles a fork well enough with the stump of his thumb.

"I wouldn't even touch a rattler that size," said the venom collector, the one they'd planned to sell it to.

————

The metaphor scientists often use is a cocktail: the venom of a rattlesnake is a cocktail of diverse toxins. There are more than thirty rattlesnake species between Canada and Argentina, and many of them have subspecies, bringing the total to over ninety. Each subspecies serves a different mix, and each snake makes individual variations on that recipe. The Mojave species packs a neurotoxin that blinds you, then makes you forget to breathe and paralyzes your heart.

Most rattler venoms break flesh down chemically. They partly digest the snake's prey before it is eaten. In fact, rattlesnake venoms evolved from digestive juices, and the poison gland of the rattlesnake is a specialized salivary gland. A good dose of venom makes your limb burn with pain as the venom digests it. Chunks of a human victim's skin and flesh may die and eventually fall off. No rattlesnake is big enough to eat a human, but the venom is strong enough to go a considerable way toward predigesting one.

Small animals usually die of shock long before the venom has softened them up, and the same can happen to a human. We can also die from such systemic effects as damage to the liver or kidneys, or from gangrene of the dead flesh.

A bigger snake is more likely to kill you because it has more venom to spend. That fact makes the diamondbacks, the biggest rattlesnakes, especially dangerous. Rattlers have personalities and moods; one rattler may crawl away while another stays to fight. Western diamondbacks generally have less patience for a human than most. They've been known to chase a man across open ground.

Snakes choose whether to waste venom on you—they can bite dry if they want to, or give you only a little venom. They need the venom for hunting, and it takes time to produce. One collector I talked to, Steve Barnum, has been bitten dozens of times. He said three-quarters of his bites proved dry. He described the symptoms of the loaded bites as "swelling, blood- and water-blisters, and a hell of a lot of pain." Most loaded bites, Barnum said, turn out to be only minor medical problems, a claim borne out by statistics: in the United States, fewer than a dozen of the thousands of people bitten by rattlesnakes in an average year die. A doctor friend told me giving rattlesnake antivenin is a risk rarely in the best interest of the patient, though the possibility of malpractice suits prompts most doctors to give it anyway. The antivenin can send a patient into a fatal anaphylactic shock.

Of course, some survivors find the bite of a rattlesnake has altered their lives.

———

At a rural construction site, one man was directing the heavy equipment by hand signals. The sound of the machinery drowned out human voices. It also drowned out the warnings of a prairie rattler that lay in the same weedy ditch where the man stood. He felt a sting on his leg, but ignored it—thorns from the weeds, he thought. He ignored it a second time, but the third sting got his attention. He saw the snake among the weeds, three feet long and thick as an axe handle. He beheaded it with a tire iron before asking his coworkers to drive him into town for treatment. It was fifteen miles.

At the hospital, the doctor presented him with a dilemma. "You can stay here until something happens, and then we can treat what happens," the doctor said. The "something" the doctor assured him was coming within twenty-four hours would present itself in the form of either a stroke or a heart attack. His blood pressure was already soaring.

"Or we can start you on a course of antivenin, which is expensive." The antivenin cost several thousand dollars per dose, and there was no telling how many doses he would need before the crisis passed.

He chose the antivenin. He passed through the usual prairie rattler bite symptoms: blood pressure rising in response to the neurotoxin, leg swelling, burning pain. He endured the treatment: an IV of glucose and painkillers, hourly blood tests, hourly checks of his blood pressure. His recovery appeared complete; his only trouble was the hospital bill, which ran into the thousands.

A few weeks after he left the hospital, he blacked out.

That blackout was only the first. He also had bouts of giddy weakness and mood swings. His doctors suspected diabetes, but after extensive tests they had to revise that hypothesis. The man was suffering from wildly fluctuating levels of blood sugar, a condition caused by the malfunction of specialized cells in the pancreas. But the problem was not conventional diabetes, and it could not be treated with insulin injections. It was unpredictable.

Now the man eats a diabetic's diet and tries to listen to his body. Probably he will never recover, but maybe he can manage his condition, which doctors call an aftereffect of his encounter with the rattlesnake.

———

Rattlesnakes like to stay in holes in the ground when they are not hunting or basking, but they're not equipped for digging. Their transitory hunting-season homes are shelters they acquire opportunistically—they simply find the place and move in, killing any occupant that objects. Rattlesnakes are often found in the former dens of prairie dogs, rabbits, and even badgers. Prairie dogs, large rodents that live in *towns* of hundreds, invented the neighborhood watch system long ago. Guards stand on their hind legs to watch and warn the others.

A prairie dog guard sees a rattlesnake coming and chirps the alarm. The adult prairie dogs defend their burrows, throwing their tails up to appear menacing and making bluff charges at the snake. They work in teams, one distracting while another rushes in for a bite. The rodents have formidable teeth and can kill a rattlesnake, though they rarely manage to. Some people say prairie dogs will seal a rattlesnake in their own burrow once he's inside, entombing him alive, Poe-style.

Sometimes prairie dogs turn the snake away from their burrows. Sometimes they don't. The snake alternates strikes and sizzling retreats, and eventually a strike lands. The prairie dog has only a few minutes to live. He convulses and dies. His comrades give up the fight. The snake retreats momentarily to save his energy while the bitten dog dies. He doesn't eat the dead prairie dog. He smells something better.

Inside the burrow, the snake finds a litter of prairie dog pups. He decides to stay for dinner. The mother comes into the litter chamber to threaten him, but he's busy eating. He buzzes a bit. She leaves.

The snake makes his living in the prairie dog town, exterminating one family and living in their burrow until he's hungry again, when he moves into the open, setting off the chirping alarms again until he conquers the next burrow. So much for the advantages of city living.

Rattlesnakes aren't picky. They like warm blood and cold, so long as it's warm enough to detect. They eat rodents of all kinds, a smorgasbord of lizards, cottontail rabbits, and birds on the ground or in the trees.

Ground squirrels frequently grace the menu. A snake learns a particular hole in the ground has good eats. He returns again and again to this same burrow, where the resident ground squirrels deliver litter after doomed litter.

———

The mating of rattlesnakes is sometimes preceded by a sort of combat dance between the males. They rear up, pressing their bodies together, knocking each other to the ground until one has had enough. Sometimes the males twine together, one pulling himself taut to send the other hurtling away, thrown like an old-fashioned top with a string. These fights, if that's what they are, also happen when females are not present, perhaps because of territorial disputes.

The male keeps his double penis inside his body, turned inside out like a pair of socks, until he is ready to mate. After he has draped himself on a willing female and done some rubbing—a sort of foreplay—one of his hemipenes extrudes itself from his vent (which one he uses is simply a matter of the happy couple's position). He inserts the spiked hemipenis into the female's vent.

The female can store the sperm for years. She may mate in succeeding years and then produce a litter sired by several fathers. In colder climates, she does nothing but eat for several short summers, building up the fat supply necessary for carrying young. She may take five years to produce a litter. The young—as few as one or more than two dozen—reside in a rudimentary kind of placenta, which is virtually an internal egg, before their mother bears them live. A mother rattlesnake does nothing in the way of child care except to strike at anyone who comes close.

Rattlesnakes are born venomous. They can already hunt for themselves. My father once reached into a patch of grass and was struck on the fingernail by a baby rattlesnake. The nail eventually blackened and fell off. He suffered no other effects. Some people claim young rattlesnakes are more toxic than adults. Possibly the explanation for this paradox is that young rattlesnakes show less restraint in using up their supplies of venom when biting defensively. A certain medical student, assuming the young harmless, handled one. He showed off for friends, telling them how ironic it is that such an emblem of fear could be handled freely. That's the way most people get bitten: an urge to handle fire. These days the young doctor has nine fingers.

Rattlesnakes are scavengers as well as predators. The

exquisite heat-sensing organs that make them ideal hunters of warm, living things also allow them to sense the recently dead. Their discrimination of temperatures is fine to the thousandths of a Fahrenheit degree. They swallow road-killed squirrels and rabbits, which lie on the shoulders of highways radiating warmth.

Like most snakes, they know many ways to move. A simple design makes them versatile. They can move in S-shaped curves, with the outer surface of each curve serving to brace the body so that it pushes forward. They can creep in a straight line by rippling the abdominal muscles. They can sidewind, using loops of their bodies as feet and essentially walking across loose sand. One desert species is named the *sidewinder* because it likes to use this trick. The sidewinder leaves tracks like those of a tank. It's not a large snake, but the sight of it thrashing epileptically across the gypsum sands, its eyes shaded by scaly horns, is enough to make most people invoke a deity or two.

Rattlers can climb trees in search of birds and their eggs. A friend of mine went fishing at a lake in north Texas and saw dozens of limbs festooned with cottonmouth moccasins and diamondback rattlers. Rattlers can squeeze into tight places and crawl on any kind of surface. They swim beautifully, holding their tails daintily above the water to keep their rattles dry. The beads need to be dry for a crisp sound.

Scientists observed a female Great Basin rattlesnake

coiling tightly in the rain. By the time the rain stopped, the groove between two coils of her body held about two inches of rain. The snake drank for half an hour from the cup of her own coils.

———

I have heard rattlesnake stories all my life. When I told my neighbors I meant to write about rattlesnakes, the stories flooded me. Many of them were obvious myth— the tree killed by a rattler's strike, the giant specimen guarding a hoard of conquistadors' gold. Others seemed plausible until I tried to trace them to the friend-of-a-friend eyewitnesses. Then they evaporated.

If you want to know how big rattlers get, you can find any length you like, up to fifty feet, in the stories. I heard as truth a story about a woman in Wyoming who shot a thirty-footer. As one researcher told me, "Snakes are like fish"—meaning the ones that get away. Their dead bodies bloat to impressive girths, and their flensed skins stretch a couple of feet beyond their living capacities, supporting extravagant claims. It's hard to guess the size of a snake, and with rattlers the danger makes it unlikely that anybody will make a point of taking an exact measurement.

Scientists draw the line at about eight feet and forty pounds, though some of them will admit a few freak diamondbacks approach nine feet. One research center has a standing offer of twenty thousand dollars to anyone who can produce a live eight-foot diamondback; they haven't had to pay. Science insists on seeing proof, which is a reasonable protection against exaggeration, but it's not a logical way to assess the upper limits of snake size. What are the odds that the biggest specimen in the world will ever run into a biologist with a camera and a tape measure? For what it's worth, one prehistoric rattlesnake species went about twelve feet.

Unlike mammals, snakes have no genetically determined size limits. They grow until they die. Their growth starts out fast but slows as they age, so that they are adding only fractions of an inch by the time they reach their twenties. Factors like nutrition and climate influence growth, but generally a big snake is an old snake. The lifespan for rattlesnakes seems to be around twenty-five years, but most rattlers don't reach the "natural" age of death. They die long before they get huge, victims of disease or enemies.

The largest rattlesnakes are those people never see. A rattlesnake risks death every time he meets a human. Some large rattlesnakes bear old scars made by human tools. Maybe these snakes, having been injured once, learned to avoid humans.

I mine the stories I hear. In the dross of fiction and exaggeration, I sometimes find the glimmer of truth:

A nine-month-old boy sat in a playpen on the lawn of his family's home. His mother and several friends were a few feet away. The boy became mildly upset about something, but he only fussed a bit, and no one checked on him. Later the family deduced that the boy must have seen the little rattlesnake at that point. A few minutes later he screamed. The adults found the snake still in the playpen; the boy had been bitten repeatedly.

He survived after a long stay in the hospital. Because of damage caused by the venom, his nervous system never developed properly. Today, as an adult, he can't drive a car, hold a job, or make change for a dollar.

———

A pet theme of nature writers and scientists is the unfair hatred humans have for snakes. It is often claimed that adults teach children to hate and fear snakes early, and that this teaching is based on their own lack of understanding. My earliest memory involving a snake is of my mother and older sister talking fearfully about a rattlesnake on a country road. We were driving and had just passed the snake. I was too small to see the road from the car and did not know what "snake" meant. I remember asking questions and getting answers that intimated revulsion but did not explain what I really wanted to know, which was what a snake looked like.

That experience fits the stereotype of learned fear, but I'm not sure it was necessarily a bad thing, except for the lack of complete information. When I was a little older, I encountered a rattlesnake in my own front yard, and fear of it kept me out of danger. My mother had no way to instruct me in the subtle differences between varieties of snakes when I was a toddler.

Another piece of snake education that has stuck in my mind came when I was about six. My sister and I were taking turns pushing each other in a wagon. She suddenly screamed for me to stop pushing. Then she leaped out of the wagon and ran away, shouting "Snake! Snake!" I didn't see the snake for a long time; it was wrapped around a post many times, its coils resembling the wrapping of a hangman's noose, its head pointing to the ground, its eyes apparently staring straight at me. That sudden feeling that I was observed, that I had been observed long before I was aware of it, gave me the creeps. My father came to investigate. He looked for a moment, pronounced the creature a bull snake, and turned to leave.

"Aren't you going to kill it?" I asked.

"He's not hurting anybody," my father said. That statement didn't make me less cautious around snakes. A classmate of mine had suffered a bull snake bite to the big toe, so I knew that *nonvenomous* didn't mean "friendly." But my father's remark did make me understand that there are several ways of seeing snakes.

The rattlesnake has served as a kind of lightning rod for human hatred of snakes. While other snakes are killed for no practical reason, the killing of rattlers has been institutionalized; their venom provides a pragmatic reason for their killing, which can easily become a pretext for killing even when other, less logical motives are the real ones.

I suspect learned fear is only part of the story. It seems to me there really is an innate fear of snakes, not only in humans but also in many other mammals.

Snakes as a group excel at scaring enemies. Cobras rear up and expand their necks into hoods; racers rush at intruders rampantly; corals have bright warning colors, and some harmless king snakes mimic the coral colors; the water moccasin flashes the fanged cottonmouth, which inspired its other common name. The bull snake coils and hisses in a warning display. Some people say the bull snake is imitating the prairie rattler; the two wear similar colors. Several snakes, including the copperhead and the bushmaster, shake their tails in leaves or grass to produce a warning buzz, and the rattlesnake has special equipment for this purpose, the rattle apparently having evolved as an enhancement of the tail-shaking behavior.

The rattler's buzz is nothing like a rattle. It is something like trickling water, and something like dry leaves on cement. It nudges my subconscious first, and then suddenly I am aware of a tickle between my shoulder blades, and I know what I'm hearing. The recognition comes fast, but I am always disturbed by the feeling that the sound was there before I heard it. This effect is universal with humans. Its cause is unknown, but perhaps resides in the ultrasonic portion of the sound.

Experiments with rats suggest another explanation. These experiments showed that sounds which provoke a reaction of fear in rats take an unusual route through the nervous system. Instead of traveling to the part of the brain that normally interprets sound, the neural message of the fear-producing sound goes directly to the limbic system. Thus, the rat reacts with fear quickly, presumably before he knows what he's reacting to. In these particular experiments, the fear response was learned—the experimenters taught the rats to associate a tone with an electric shock. The human reaction to the rattlesnake's buzz does not have to be learned. People who have never seen a rattlesnake get the same tickling sensation in the back of the neck on first hearing the buzz. The reaction occurs when the buzz is used in recorded music and isn't even consciously recognized. This reaction suggests that our fear of the rattler is instinctive, perhaps ingrained through long generations of human, and prehuman, danger. However, the human race came to the Americas, where rattlesnakes are found, only about ten thousand years ago. It may be that the fear reaction is even more universal than it first appears, a fundamental aspect of having a mammalian nervous system.

Add to the snakes' own artistry our fear of anything different—the snake eats and makes love and shelters himself from the cold like we do, but he moves without legs, like something purely hungry, purely sexual. My accountant and I were swapping rattlesnake stories one afternoon, and I mentioned the idea that we hate and fear them by instinct.

"Ever since the Garden," he said.

———

Venom makes its user a specialist. Whereas more primitive snakes simply swallow prey, or else suffocate it by constriction, venomous snakes have the option of delivering a killing strike and then allowing their prey to die before moving in for a meal. This tactic spares them some of the danger involved in overpowering prey; living prey can bite and claw.

The venomous bite has evolved independently in many different animals, from octopi to shrews. It even evolved independently in different types of snakes. No one knows exactly how it came about in rattlesnakes, but some clues can be found in the behavior of the monitor lizards.

The monitors constitute the lizard family most closely related to snakes. This relationship is not hard to spot once you've seen a large monitor move: it walks on legs, but with an ophidian essing of the body. In the largest monitor, the Komodo dragon, the males engage in courtship battles similar to those of rattlesnakes, rising into the air as they push against each other in a sort of sumo match. And, most snakelike of all, the Komodo dragon smells by constantly lashing the air with its forked, black tongue.

The big lizard (up to about ten feet and three hundred pounds) can find fresh carrion more than a mile away by scent. Its sense of smell is, in fact, more acute than that of a bloodhound, which itself can seem almost supernatural to us. But the dragon doesn't restrict its diet to carrion; it also actively hunts, and its predatory technique is strikingly similar to that of the rattlesnake. The dragon rushes the prey—which can be something as large as a pig, a deer, or a human child—and delivers a toothy bite. Then it allows the prey to escape, and tracks it by scent. The dragon has no venom; the tactic works because of the festering meat between the reptile's teeth, which makes the bite septic. In the hot, humid tropics, a septic bite can kill a big animal in a couple of hours.

The rattlesnake's venom may be a refinement of a septic-bite tactic, which both the monitors and the rattlesnake's ancestors might have developed to complement their extraordinary smelling abilities—an example of convergent evolution. Killing by septic bite is also a feline tactic. The lynx uses it on young caribou. Many people mauled by lions have died from wounds that should have been survivable: the meat caked under the attackers' claws and teeth injected the victims with disease, and they died in a gangrenous fever.

Evolutionary theory suggests that the rattlesnake's

venom would improve its chances of survival. Reality, however, proves more complicated than theory. While the venom enhances the rattler's success as a hunter, it also creates unique survival problems.

In the American Southwest, there's a tradition of killing rattlesnakes as a point of etiquette. It goes back at least to the middle of the nineteenth century. The idea was to kill any rattler you found, even if it wasn't threatening you, so that it couldn't bite somebody else another day. Farmers and cowboys would decapitate rattlesnakes with a whip or a lariat, striking from a safe distance. Nowadays people more often kill rattlers by running over them on the road.

Rattlesnake hunts or roundups are a ritualized, and commercialized, version of this old custom. Roundups reportedly began as a way for farmers to make their fields safer to work in. The tradition began in colonial times, with an entire community setting aside a day to catch or kill every rattlesnake they could. The snakes gathered were boiled to produce snake oil, one of the infamous "medicines" sold by traveling hucksters. The Okeene, Oklahoma, chamber of commerce boasts the first roundup in the modern sense; essentially, it's a tourist attraction. Okeene and other towns, like Sharon Springs, Kansas, and Sweetwater, Texas, hold annual festivals featuring such amusements as measure-offs, cookouts, bagging competitions, flea markets, dancing shows, carnivals, and exhibitions of bravery such as peo-

ple sleeping in snake cages. There's something archetypal in these events: they're like pagan celebrations of
spring. In one town the roundup is always scheduled to
follow Easter by a week.

There are no festivals for nonvenomous snakes, and
no point of etiquette in killing them. The rattlesnake's
venom makes it a target for predators—in this case,
humans.

————

Recall the smell of a restaurant that serves fresh lobster
but isn't scrupulous about cleaning the lobster tank.
Now, in your imagination, dry this smell out, and spice
it with the dried rust of old water pipes. For the sound,
imagine being in a shower, the water pressure rising
and falling erratically.

This is the Rattlesnake Pit.

The so-called pit is not underground. It is a fence of
welded iron and lumber set on the cement floor of a
community building in Waynoka, Oklahoma, where
the roundup is under way. The fence is five feet high
and runs about twenty feet by ten, and inside it at this
moment are 203 rattlesnakes. Two men work the pit.
They use metal rods to unstack the snakes in the corners, where they tend to accumulate. Unstacking is necessary because even though these snakes can't climb a
smooth five-foot perpendicular, they might be able to
use each other's bodies as steps to shorten that distance.

The pit contains prairie rattlers and Western diamondbacks. Each year the longest prairie rattler caught in this roundup runs about four and a half feet, and none of the prairie rattlers in the pit is up to that mark. They are slender, though their girth increases slightly from neck to near the tail. The longest diamondbacks listed in the roundup records exceed seven feet. Of the ones in the pit, only a few approach six feet. The hunters keep their best catches of both species until the measure-off; the pit is full of their rejects.

One of the men in the pit, a powerfully built, sloping three-hundred-pounder, uses a special metal rod—a snake hook—to pin the head of a big diamondback to the cement floor. He seizes the snake, index finger atop the head, thumb opposing the other fingers at the sides of its neck, and picks it up. He brings it around to show everybody, face to face. People are packed two or three deep around the pit, children holding each other up to see, men suddenly pinching women from behind to make them think they've been bitten, the mass of people smelling of whiskey and cotton candy and sweat and Marlboros—but the snakes smell stronger.

The handler holds the snake's open mouth two inches from my camera lens. The hooked fangs drip venom that looks like cloudy white wine. The smaller teeth that lie in paired stripes on top and bottom are not visible; all I can see of them is the holes through which they have retracted. The holes are swollen and red, probably from the handler's having pushed the snake's head against the floor to catch him.

"How many times you been bit?" a spectator asks.

"A few," the big handler says.

He deflects most of the questions in this way until someone says, "Are you ever afraid one of 'em will reach up and bite you in the balls?" Then he becomes animated.

"Only me and the gal that does my laundry know how much."

The other handler comes by a little later, kicking a pile of snakes out of the corner. They flare up at him, raising their heads shaped like the spades on poker cards, pulling back along the looped lengths of their own bodies to strike, their noise increasing as the hiss of a fire does when you throw on gasoline. The man is wearing heavy boots, and most of the strikes bounce off. One big diamondback hangs a fang in the man's jeans, and man and snake have to wrestle a moment to be free of each other. Later, another snake strikes the man's boot and leaves a half-teaspoon of venom on the scuffed leather. "He was pumped up, wasn't he?" the man says.

A third handler enters the pit. He catches a six-foot diamondback and holds it by the tail; it lifts its head, making its body a parabola. Its black forked tongue flits and gleams. The handler gives it to a girl perhaps ten years old standing outside the pit. She holds it by the tail for a minute before handing it back. Now everyone is reaching for it. The handler shows them how to grasp it—thumb up, rattle up, at arm's length, so that the snake can't double himself against gravity and score a bite. "He's heavy," a woman says. The handler takes the snake back and considers.

"About a seventeen-pounder," he says.

At the other side of the pit a man appears with a gunnysack. A handler lifts diamondbacks into the sack with a snake hook. The man with the sack goes to a back room. I follow for the butchering.

———

The contents of the gunnysack are dumped into a fifty-five-gallon barrel, which echoes metallically with the buzzing of the snakes. One man uses a mechanical grabber to seize a diamondback by the head. He holds the head down on the chopping block, never letting his artificial grip loosen.

A second man grabs the tail with his bare hands. They straighten the snake, which fights them. They slap at its curves to make it relax. A third man measures about six inches back from the head and slams his hatchet home. The snake writhes, both its neck and its headless body twisting. The man with the grabber drops the head into a green plastic bucket already half full of heads. The hatchet man points the wounded body into a barrel to let it bleed out. The blood stops in less than a minute.

The hatchet man sets down his hatchet and takes up a filleting knife that hangs from a chain attached to his belt loop. After slapping the headless, twisting body into submission again, he makes one long incision from the neck wound to the vent. He peels the skin back from the neck; it comes off easily, like the skin of a banana. He stops skinning at the tail and chops; the tail and skin go into another bucket. He reaches into the long incision he has made and pulls out the organs, along with a string of yellow globules of fat. They are all neatly contained in a membrane that pulls out easily.

What's left is a long sleeve of muscle and bone. It still writhes a little. A woman takes the sleeve and puts it into a sink full of soapy water. Its writhing accelerates until it is rising out of the sink in great hoops. The woman seems unperturbed. She washes the outside and the empty body cavity and then sets the sleeve in a sink of rinse water.

Another team takes it from there. They stretch it out on a marble-topped table and use a meat cleaver to cut it into fist-sized chunks.

Volunteers sell the meat in plastic bags, ten dollars a pound. The other parts go to an artisan in Colorado who will freeze-dry the heads to sell as novelties. The rattles will become key chains and such. The skin he tans and uses for hatbands and belts and wallets. The leather is delicate, grained with scales the shape of sunflower seeds, and far less practical than the kind made from the hides of cattle. It is only the snake's reputation as a killer that creates a demand for its body parts.

Similarly, people eat rattlesnake meat because the creature has such a glamour about it, such a reputation for dealing death. The meat is expensive and not especially easy to get, yet it attracts people who would not care for other high-priced foods—snails, for example, or ostrich.

I myself have eaten rattlesnake. I bought a couple of pounds at the Waynoka roundup. A rattlesnake hunter from Texas told me the best way to cook rattlesnake meat is to brush it with lemon and butter and grill it, though people more commonly bread it and panfry it. This enthusiastic snake-eater said the taste had "a wild richness." An acquaintance who grew up in Waynoka told me how he had been "uninvited" to a neighbor's barbecue after he announced his intention to bring rattlesnake. This gentleman described the taste as resembling that of breast of chicken, but I put little stock in this claim, since I have heard every meat from frog to alligator similarly described. I have eaten frog, alligator, and a number of other supposed chicken substitutes, and none of them impressed me as particularly chicken-like.

Following the Texan's recipe, I basted the rattlesnake meat and put it on the grill. Thirty seconds later each chunk of meat had shriveled to the size of a silver dollar and was the texture of a steel-belted radial.

"They look like little vampire bats," my wife, Tracy, said. She was right. I tied one to a string and caused it to fly around the backyard while I made squeaking noises. My son, who had just turned one, thought it was funny, but Tracy gave me one of her patented see-what-I-have-to-put-up-with looks. Then she put some chicken on to grill.

"Surely you're not giving up on the snake yet?" I said.

"There's no meat on your bats," she said. She was right again. Each little bat consisted of many tiny ribs. Between any two ribs there was about enough meat to make a mouse need a toothpick.

Nevertheless, I set about eating. The taste wasn't bad—something like gamey, dark turkey meat, and, to be honest, a little like chicken—but I nearly starved trying to get the meat off the bone. After the meal, I phoned the Texan.

"What sort of scam are you trying to run?" I asked him tactfully.

"It always worked for me," he said, as if he'd done nothing wrong.

Despite my disappointment with my own culinary skills, eating the rattlesnake gave me a deep satisfaction. The taste was not the point, just as a human cannibal's point in eating his enemy's heart is not gustatory pleasure. It is the devouring of death that matters. It is a communion.

Rattlesnakes live in a world of vibration and scent and subtleties of heat. Dogs live in a world that's largely scent and high sound. Somewhere in that system of senses beyond our experience, a dialogue of hatred transpires.

When I was a child my grandparents had a German shepherd. I saw her one day moping around the porch, her neck swollen as if with a goiter. My grandmother told me she must have fought a rattlesnake. That's the first I knew of the ancestral feud.

One morning our own dog, a miniature bloodhound, lay under the air conditioner. He had arranged straw and an old blanket into a sow's nest in the dirt, as if he expected to be laid up awhile. He growled when anyone approached. He had never growled at us that way before, with that tone of serious business. He lay there for two days, and the bowls of milk and water we set at a safe distance went untouched. They were cold, drizzling days, and the air on his side of the house smelled like soggy toast and sickness. My mother spoke to us gently, making implications I didn't want to hear.

On the third morning he was up and surly, snapping at the other dogs, licking weakly at a pan of gelid milk. Loose skin sagged at his throat, the only remnant of the throat-constricting bloat I had never quite glimpsed. I had only heard my father tell my mother about it.

I assumed the dogs must have been taken by surprise, but that's not how it works. Dogs like to hunt rattlesnakes. They detect the scent easily. There's a confrontation—even cowardly dogs seem to lose all sense in their hatred of the rattlesnake. As the snake becomes more agitated, so does the dog. I suspect the sound produced by the snake's rattle, which affects people so profoundly, does the same to dogs. Or maybe it's the stink of the rattler. Some claim the smell can make a human dizzy.

If a dog gets his jaws around a rattler, the rattler is doomed: the dog whips its head side to side, snapping the snake's spine, and doesn't stop until the snake lies in pieces. (Alligators kill rattlesnakes by the same method, which they use only on venomous snakes—the others they simply swallow.)

Dogs lead with their snouts, so they almost always get bitten somewhere on the front end. Some die of the bite, though in general dogs are fairly resistant. Some dogs hunt a rattlesnake, once scented, to the exclusion of everything else. One farmer claimed he could tell from a distance when his dog tangled with a rattlesnake by its tone of voice.

Dogs become more efficient at hunting rattlers through experience, and experience involves bites. The canines develop immunity to the venom, and some old dogs hardly notice another rattler bite. The first bite makes the dog, which innately dislikes rattlers, hate and actively hunt them. Maybe the nasty near-death experience amplifies their hatred, or maybe they somehow know they've been immunized and can indulge their passion freely.

———————

Estimates of the rattler's size ranged from eighteen inches to four feet. Personally, I put it near the smaller end. The question is not trivial. My boots are thirteen inches high—I measured them afterwards—and knowing the snake's length would theoretically have allowed me to judge how close I could get, because rattlesnakes are said to have a strike less than half a body-length. As it was, I chose to keep a good distance, and my wife said later that while I was at work with the shovel, I formed an arch like a pole vaulter in midleap.

We were inside the trailer house when the dog raised the alarm. Jody, my brother-in-law, opened the front door to investigate. I was sitting at the kitchen table in front of the swamp cooler with a glass of iced tea and couldn't see outside, but I didn't have to. I heard the buzz.

"Another one," Jody said. "Well, hell."

"He hates those rattlers," Corey said.

I asked if she thought her husband would take offense at an offer of help. She thought he'd be grateful.

I asked Jody for a shovel. He brought one for himself as well. Everyone else stayed in the trailer.

The prairie rattler was a good twenty feet from the door, but its buzzing was audible over the dog's barking. The dog, a young chow, worked a circle just beyond striking range. The snake coiled back and back on itself, its head leveled for a strike, its tail shivering too fast to be visible.

Jody and I approached from opposite sides. The dog fell in behind me and kept barking. No matter which way the snake and I moved, the dog's nose remained pointed to the true-north of the rattler.

I planted the shovel blade in front of the snake. Sometimes a rattler will stand still when faced with a shovel blade. If the metal is cold, it confuses him—he senses it as something moving but too cool to be alive. I know a man who collects live specimens by employing this fact. He puts a shovel in front of the snake, and while it pauses to "look" with its heat sensors, he walks up behind the rattler and picks it up by the back of the head.

But this day was well over a hundred degrees, and I guess nothing read "cold." The rattler lunged up the shovel blade, and kept trying to climb it as I wagged it from side to side. It was zigzagging, but its movements looked like those of rapid water. It could have climbed the shovel easily if I'd held it still. Instead, I flicked the snake back and slammed the blade down on it. The stroke wasn't clean, but it was the best I could do with a rattler stirred to such speed. The rattler rolled over twice around the wound, which cut across two of its coils. I saw the white flash of its belly, then the dusty gold and chocolate of its back, then both again in quick succession. The buzzing didn't stop.

All that movement happened in a second or so. I planted another blow, and this time, with the snake wounded, my aim was better, the blow severing the head cleanly. The buzzing stopped, then started again in a muted way, as if the tail had plunged under water. The body whipped once, then straightened out; the muscles still moved, but forward, in a sort of peristalsis. I drove the shovel blade into the ground with my boot to make the decapitation sure. The buzzing gave a few stops and starts—and then the tail stilled. The dog leaped forward, bit the snake near the middle, and slung the body side to side.

The dog carried the body off somewhere. I took the rattle. Jody buried the head to prevent anyone's stepping on a fang.

————

That was the third rattler they'd seen around the trailer. A couple of days after we killed it, Jody and I walked through the tousled grass behind the trailer, talking out the problem.

A small frame house had stood in what was now the front yard. The house had stood empty for twenty years before Jody had had it bulldozed into its own cellar. It was completely invisible under the yard, which Jody and I had tilled and planted with grass seed. Occasionally as we tilled we found an old shingle or a shard of window glass. The only outbuilding left was a barn. There had been no cellar to push the barn into; the workers had simply knocked it over. It lay in untidy piles in what Jody and his family had designated the backyard, its wood weathered past any trace of paint. Yellow grass jutted between the boards. Three parched elms, leaning with the course of fifty years' wind, stood around the pile. They were the only trees within eyes' reach.

The fallen barn was clearly the source of the snake problem. It was good shelter, inaccessible to hawks and dogs. I poked around it. In places the dirt gave way beneath my boots; the bulldozer had leveled the ground but left hollows just beneath the surface. Once I dropped about a foot. The crater I had discovered was a sort of cave extending at least six feet to the side. I scrambled out, fearing the cave was a rattlesnake den. There was a jumble of pipes in it that neither of us could explain, and a mass of concrete that must have been part of the old house's foundation, but no sign of rattlesnake.

I went farther into the rubble. Jody's two-year-old son followed me. When I realized he had come along, I said, "Go back to the house—there may be snakes in here." I hadn't finished my sentence before I heard a buzz like the shivering of icy leaves.

"Get back," I said to the boy, in a tone I hoped would discourage any questions.

"Why?" he said.

Jody came running. He had heard the buzz. He scooped up the boy and ran him to the trailer.

"Bring a shovel when you come back," I said.

So far I hadn't seen the rattler—I had only heard it. Now the buzzing stopped and I glimpsed movement. A fallen outer wall of the barn lay six feet to my right, and a thin snake was fleeing along its siding, smooth as oiled rope through a pulley. Its hide had the faded look of the prairie rattler, the yellow of old documents patterned with the brown of old saddle leather. Its head slipped into a hole in the wood, and the body glided after it without any break in its rhythm.

I had lost time seeing to the boy. My chances of killing the snake were slim. I saw one opportunity: the board the snake had slipped under looked loose. I kicked it. The bent nails that held it creaked and the board flew off.

The snake was holding the same straight line it had been following on top of the wall. A man who drives a ditcher once told me of following a rattler in a new ditch. The man had dug the first pass and needed to make a second pass to deepen the ditch. He saw the snake crawl into the ditch behind him on the first pass;

when he turned around to make the second pass, the snake fled along the ditch for dozens of yards before being flung by the ditcher. It could have saved itself by crawling out of the ditch. They are like water: path of least resistance. They don't see things schematically, from above, as we erect primates do.

My snake was already halfway under another board. Its rear half was exposed. If I had had a shovel, I could have delivered a deathblow with little danger to myself. As it was, my only chance of killing it was to grab the thing in the second before it made cover. The only safe way to hold a rattler is by the back of its head, but if I could grab this one by the tail and throw it onto open ground, I might be able to block its escape until a weapon came to hand.

I reached for it—and glimpsed something disturbing. I hesitated, and the snake escaped.

I had glimpsed a tangle of spiderweb. It glimmered in the shadow of a board as the snake brushed by. I had seen only that glimmer, but it was enough to suggest the shape. It was a black widow's web.

Jody returned with a pair of shovels. "How about a jar?" I said.

"You're not going to fit that snake in a jar," he said.

After I had explained the new development, Jody procured a Miracle Whip jar. I looked around thoroughly to make sure no snakes were likely to bite me while I knelt over the widow's web. The web ran in a sloppy weave along a board. Beneath it I saw the glinting black shell and pincers of a carabid beetle; it must have been a massive specimen, at least two inches long. The other insect husks, covered in mold and old web, were harder to identify.

The web twisted into a kind of funnel which angled under the board. Because of the turmoil the snake and I had produced, the spider was almost certain to be hiding in its funnellike retreat. I turned the board and slapped the jar over the funnel. A flick of a twig knocked the spider into the jar.

It was a mature female. Despite its dirty surroundings, it gleamed a clean black. I put the lid on the jar. The widow scratched frantically at the glass, but couldn't climb it.

In the widow's web hung a marble-sized ball of silk, its texture like linen. It was an egg sac, and in the side of it was a pinpoint hole through which the spiderlings had already escaped. The egg sac suggested the possibility that young widows might be scattered throughout the woodpile. I stood and thought about the situation while Jody went for the dog. I hated to let the snake go; it was too close to the trailer for safety. I also hated the idea of digging through the wood; I'd be in constant danger from black widows and rattlers both. The black widow has a more dangerous venom than any rattlesnake ounce-for-ounce, but the rattler deals in greater volume. Both venoms are less dangerous to dogs than to humans.

Jody brought the red-blond chow from her pen. Her black ears had serrated edges where they had been fly-bitten. She barked wildly at the woodpile. She had caught the snake's scent. I could smell it too, a clammy, nauseous taint in the air. She went to a spot in the middle of the fallen wall and tried to dig, her claws rebounding from the wood. She didn't make a dent, but she kept digging. I came over to stand on the spot; she backed off to let me have my turn. I stomped on the boards. After five or six tries, my boot broke through to dirt. I stepped away quickly. The dog plunged in, weaseling through the hole I had made and digging at the ground. Dirt flew out along her red flanks. Her barks turned to growls. She had something.

She came out of the hole tossing her head from side to side, the snake swinging in her mouth—but it wasn't the snake. It was only his old skin.

———

The next afternoon was 110 degrees. It was lousy weather for a snake hunt; no snake would be out in such heat. It would be deep beneath the old wood until dusk, pouring its excess heat into the cool dirt. But this was the only time I had, and I went at the pile with an axe Corey found for me. I meant to chop through the wood and stir something up.

I brought the dog with me. She barked at the smell of the place and scratched at the hole where she'd dug out the old skin. She dug at the loose boards where I had seen the snake the night before. I worked with the axe, but it was dull. I didn't get far before the heat wore me down. I stood in the hot wind with red wasps buzzing around my face and the dry grass hissing—it sounded a little like a rattler's buzz, and I confess I had stopped in my chopping several times to make sure of the sound.

I couldn't do much more. It was time for drastic measures.

———

The bulldozer returned to push the wood and its tangle of grass away from the trailer. It gouged a hollow near the old house's grave, and then scraped the hollow again, pass after pass, and soon the hole was deep enough to hold the barn. The dozer left the land smooth, and where the old farmhouse and its outbuildings had stood for seventy years, there stood only a trailer house and a patch of newly sprouted grass.

———

A couple of weeks after the bulldozing, a man from the county came to tell Jody and Corey to be on the lookout for snakes. The county workers were mowing the ditches of the dirt road near their house, and whatever wildlife lived in the ditch might show up at the trailer. But the family saw nothing unusual.

Driving away from their place one hot evening, I spotted a rattler on the road, recently dead from a sharp wound, maybe the slash of a mower blade. Sun and rot had not yet made it swell. I took the head and the rattle.

———

You've had nightmares like this. Everything is normal except for one detail, preposterously wrong, and no one else is there to notice that vulgar disproportion, that dangerous but unspoken violation. The young man was looking across a long stretch of ground at three horses watering at a trough. Two of them seemed themselves, and the third had an enormous head. The young man stared, trying to see things in some different way that would make some sense of the scene.

He approached the horses. He had worked with all of them, and they hardly stirred at his approach. The two buckskins drank, water rattling off their tongues. The big palomino mare turned her rich brown eyes on him, but they were no longer horse's eyes. They were bulging and weeping blood. Her head was twice the size it should have been, and furrows of creased flesh showed where a halter should have been. Her breath was coming hard.

The man touched the horse's face. He should have felt the firm bone beneath the felt hide, but her muzzle was softer, like a bag of water. He ran his fingers into the furrows and found the straps of the halter buried there. He fumbled to unbuckle the halter, his every tug on the straps causing a catch in the mare's breath. Finally he took out his knife and cut the halter free. The mare breathed easier and plodded to the trough.

He'd seen the two punctures just above her nose. They told him everything: a patch of grass or weeds, a warning buzz, a young grazing horse too curious to heed it.

———

Scientists think the rattle evolved as a way to warn off hoofed animals that might accidentally trample the snake. Horses seem to understand the message, at least when they're older and experienced on open range. They usually shy from the sound.

But there's more to the story. Observers have seen horses go out of their way to trample rattlesnakes. Other hoofed herbivores—pronghorn antelope, deer, cattle, sheep, goats—also actively attack rattlesnakes. One deer was spotted jumping up and down on a rattler for half an hour. No other small animal gets this kind of treatment from the hoofed contingent. Why do large herbivores attack this particular snake? Maybe they, like some people, want to kill the snake while it's in the open and the odds are good, somehow knowing they may meet it under worse circumstances another day if the snake escapes. The rattlesnake's venom, a prime predatory asset, brings down its doom.

———

A man rode over grassy pasture toward a stand of oaks, where he hoped to have his lunch in shade. He noticed a bird weaving among the trees while he was still a mile away.

He thought the bird's odd flight might make more sense as he got closer. Rushing up from somewhere near the ground, spewing some raucous invective he'd never heard from an avian throat, turning up rapidly and then spinning into a kamikaze dive, the bird was so thoroughly occupied that it didn't react to the man's approach.

When he was within the shade of the oaks, the man smelled rattlesnake. His horse shied a bit. The snake was moving out of the grove, rattling sporadically. The bird, rain-cloud gray with bursts of white, was dive-bombing the rattler. The man looked carefully—as carefully as he could without dismounting. The snake was wounded around the neck, and its bloody face was eyeless. The bird visited a nest in one of the trees, then dove again: a rush of threatening screams that only a mockingbird could make, a rising buzz, and then the bird rose, the whole pass too fast for the man to see what exactly the bird had done. The snake must long since have given up trying to strike the faster bird. He simply dragged his ragged, dying body toward the tall grass.

Mockingbirds are territorial, and will attack even people and dogs that venture near their nests. They've been known to keep after a rattlesnake for an hour. They don't relent even when the snake leaves their territory; they follow and perform an execution.

This all depends on the bird's detecting the snake. Things can go the other way if the snake gets in the first shot. Rattlesnakes have been found in the nests of mockingbirds, eating every last egg.

The extravagant violence between reptile and bird makes a more obvious kind of sense than the preemptive killings by hoof. Rattlesnake and mockingbird are natural enemies whose relation hinges on predation, the snake trying to eat the bird's young, the bird getting nasty at first sight of the snake.

If the mockingbird reacts to the rattlesnake with violence born of fear, several other birds react to it with fearless hunger.

The golden eagle plucks a rattler from a vast expanse of wind-rippled grass. You see the silhouette rise into the sky: the bird's wings slapping the air like sheets on the line, the snake twisting and knotting in the rugged talons that have already dealt him fatal wounds. Sometimes the snake manages to reach the eagle with a fatal strike as they struggle high in the air.

The rattler is lethal at one end and scary at the other,

but in between it's a tube of protein irresistible to many predators. Hawks and owls take rattlers, but so do some less obvious avian predators, like wild turkeys and domestic chickens. The tough, scaly legs common in birds give them an edge over snakes. The roadrunner specializes in rattlers and never seems to get bitten.

Other rattlesnake predators include domestic cats and pigs, skunks, badgers, bobcats, coyotes, foxes, and other snakes. Some kinds of kingsnake live mainly on rattlers, which they kill by constriction and swallow whole. There's a beautiful iridescent pink snake with a texture like braided rawhide that hunts by sight, rearing up and looking around for prey. It's called the *coachwhip,* and it often eats rattlesnakes.

Whatever advantages the rattler's venom may provide, they don't include freedom from predation. To some animals, the rattler is only a potential meal; to an astounding number of others, it is something so fearsome it must be either fled from—or killed on sight. No other animal provokes such visceral reactions from other species.

———

A couple rented a building for their motorcycle accessory business. The poured cement floor of the old building was divided into neat islands full of chrome and leather. Several months after the shop opened, the weather turned cool and the trouble started.

The woman was sitting at the desk doing paperwork when she heard a sound she described as "someone turning on a shower." She pushed away from the desk, her swivel chair shooting back on its rollers. In the instant of moving she looked down to see the rattlesnake's strike in a blur. The strike fell short of her retreating leg. About an inch short.

The woman ran for her baseball bat. She claims the rattler then became the recipient of a twenty-minute batting practice from which it did not emerge alive.

Up to this point, her experience was not unusual. Rattlesnakes do occasionally come into buildings for warmth. For example, some friends of mine had a house in the country. They came home one evening to find a sizable rattler coiled beside the washing machine in the utility room. The gentleman of the house prodded the intruder outside with a broom before demolishing it with buckshot. My friends had a situation ideal for drawing rattlesnakes. The utility room was flush with the ground. It had, besides the usual warmth of a human habitation, heat-producing machines (washer, dryer, and water heater) that would have been "visible" to the heat sense of a rattlesnake at night. And its floor was cement.

A surface of cement might as well be a rock for the rattlesnake's purposes. Like flat rocks, cement surfaces soak up the heat of direct sunlight and pour that heat back into the air at night. Rattlesnakes use such surfaces to control their own body temperature. For example, a rattler emerging from his hiding place at dusk may lie on a radiating rock to get his blood warm for the hunt.

To my friends, the cement floor in their utility room was a clear indicator of a human territory. In fact, they seemed to consider the rattlesnake's trespass not merely a danger but an insult. The snake, however, probably wasn't even aware it would encounter humans, if in fact it had any concept of humans. It simply got cold and followed the heat across a surface that seemed natural enough and easy to move on. It must have had to squeeze through a crack or under the door, but snakes are good at that.

This problem, disconcerting as it was for my friends, was minor compared to the difficulties encountered by the couple at the motorcycle accessory shop. The day after the close call with the rattlesnake, they discovered a garter snake in the building. The following day brought a kingsnake, and the day after that a bull snake. When I came to visit the shop, no one was there. The cement floor of the building was flush with the ground, and there were several holes in the surrounding earth. A sign on the door read CLOSED DUE TO SNAKES. When I contacted the proprietors, they seemed ashamed, as if they had been publicly exposed as lepers. They found a new building for their business.

The low cement floor had allowed a few snakes to get inside, but the indoors wasn't their destination.

What drew them in the first place was surely the presence of an ancestral den beneath the building's foundation.

————

When a rattlesnake eats another animal, he eats it all—not just the muscle tissue and organs, but also the bones and fur or feathers. Each prey item he takes is likely to be huge in proportion to his own weight: a four-pound snake eats a two-pound rabbit as readily as a man eats a quarter-pound hamburger. Once he's converted the meal to energy, the snake doesn't spend that energy keeping his body temperature constant, as we mammals do. All of this means the snake can stretch its meals a long way. In a laboratory experiment, one rattler survived a year between meals.

Rattlesnakes adapt to different climates by exploiting this ability. They operate most efficiently in a hot climate, producing abundant venom and hunting year-round, but they can endure winters by simply crawling into holes and hibernating. These hibernating places are dens. The difference between a den and a temporary shelter used in spring and summer is that a den houses an entire congregation of snakes. Denning is the weirdest thing about rattlesnakes. Hibernation doesn't fully explain it, but it's a place to start.

A den can be any cavity—a sinkhole, a cave, a man-made well. Heat helps determine the site. Dens are near flat rocks, ledges, or open ground; rattlesnakes coming out of hibernation need to bask in sun. In the North, dens turn up in the crevices of south-facing slopes—southern exposures get more sun. In the South, where winters are mild, a dip in the ground will do. A friend of mine was reminded of this fact one day while he was hunting. He stepped over a fallen tree onto what he thought was solid ground. It was actually only a tangle of twigs and leaves that covered a waist-high depression. When he fell in, he heard the sibilant greeting of many rattlers. He quickly arranged to view them from a distance.

In Texas County, Oklahoma, where I live, the land is mostly flat and free of holes and cracks. The prairie rattlers here use the only available holes, the same ones they use in the active season: prairie dog burrows. One particular prairie dog town in this county extends for something like three square miles, and some of the burrows there house several dozen rattlers each in the winter.

The scarcity of holes in this landscape has an odd side effect: the group hibernation of natural enemies. The rattler sleeps with the bull snake and the blue racer, species that in their active phases prey on rattlesnakes, killing them by constriction or simply seizing them by the head and swallowing. The prairie dog and the pack rat, perennial victims of snakes, nestle into ophidian masses for the winter. The rattler lies down with its own predators, the badger and the fox.

A few hours east of here, there's a country of red earth dominated by mesas. Centuries of rain have carved arroyos down the sides of the mesas. That's where the Western diamondbacks live, in those red ravines: the country with bigger caves has bigger snakes.

Denning in numbers helps conserve body heat. The rattlers need to hold a little warmth even when the ground is frozen deep. You can find dens along creek beds where leaves accumulate and rot in the cave-pocked banks, the heat of decomposition warming the snakes. My father worked in a feedlot, where thousands of fattening cattle were packed into pens, eating from a trough, the ground beneath them covered in their own manure. The dung would build into great hills; I often saw mounds taller than the fences. My dad's job was to whittle down these hills with a front-end loader. He unearthed tangles of snakes beneath the dung: bull snakes, blue racers, and prairie rattlers basking in the subtle rot-heat. Once he dug up a globular snake-mass a foot and a half thick.

In north Texas a house had stood vacant in the country for years, and when electricians showed up one day to reclaim it for human use, they found it claimed for denning. The men killed eighty snakes and stacked them like dirty laundry.

———————

After a winter in the den, each snake goes its own way to hunt and maybe mate. A rattlesnake returns to its original den with the coming of cold weather. Newborn rattlesnakes winter in their mothers' dens. These young snakes have never been there before, but somehow they find their way to the ancestral den. How they do this, and how adults find their way back, is a mystery. Rattlesnakes may navigate by the sun, or they may memorize the way landmarks look to their heat-sense, but these mechanisms don't explain how the young find a place they've never been. One researcher who had spent a great deal of time tracking rattlers, even cutting into the bellies of some to plant radio transmitters before sewing them up and releasing them, threw up his hands and blamed "instinct."

Another theory is that the young track the adults to the den by scent. In laboratory tests, young rattlesnakes followed the paths of adults in simple mazes. Whether scent is involved in the wild, no one knows. The young rattlers may be miles from the den and may not have seen their mother for months; if they track only by scent under these circumstances, it's an astounding feat.

But the chemical senses of the rattlesnake are so much better than ours that our idea of smelling is a shadow of theirs; we are like congenitally blind people who see only vague masses of gray and light and cannot

grasp a sighted person's feelings toward blood, dia-
monds, autumn, and Renoir. The rattlesnake can
identify the scent of any of his den mates as family.
With dozens or hundreds or thousands of summer-
wandering snakes trailing the taste of home, the young
rattlesnake perceives a world reticulated with traceable
connections, a network that reads as clear as the spider's
web does to the touch of its maker or a system of highways does to a human.

———

When rattlesnakes convene for denning, they first form
a bolus—a ball-shaped cluster, like a collection of rubber bands. Every member of the bolus keeps moving,
the pulsing amalgam growing as more snakes arrive.
One man peered into a cave and saw a bolus more than
four feet thick. There are bigger claims, too, if you want
to believe them.

Writer J. Frank Dobie reported the story of a hired
man sent to bring in two grazing mules. The man's boss
heard a scream, and then a fainter one. He found the
body in a gully amid hundreds of rattlers. The snakes
were forming a bolus. The man, who must have
stepped into the gully without looking, was already
dead.

A bolus soon breaks up, and the snakes enter the den.
They may form the bolus as a ritual of recognition. Or
maybe it has some obscure connection to mating. Or

maybe it's a way of scent-marking each other to help with next year's den-finding. Nobody really knows.

The number of snakes in a den varies with species, climate, and predation. Some dens contain only three or four rattlers. And then there are the big ones.

———

My friend Michael Gabriel was hunting dove with a friend on a ranch in the Texas Panhandle. Gabriel had a bird in sight. It was flitting between a low cottonwood branch and the edge of a puddle. It was close enough for Gabriel to see the plump muscles of its gray-brown breast. He was moving closer, trying to be unobtrusive. He could hear the bubbling coo of the dove—and then, cutting through that smooth sound, the buzz of a rattlesnake.

He looked around until he spotted the diamondback lying on the ground three yards away, a safe enough distance for him to stay and watch. The other hunter, having heard the buzz, cautiously joined him. The snake had stopped rattling. It crawled off at a leisurely pace, though not exactly away from the men; instead it was heading toward a nearby tailwater pit. It buzzed again, as if in afterthought.

Behind the hunters, another buzz answered. They looked around, more concerned this time. They spotted the new rattler, and as they kept glancing around they noticed other rattlesnakes, invisible to the casual glance, but suddenly popping out of the landscape. There were about a dozen in sight, and they stirred and moved, all of them headed for the big tailwater pit. The air was suddenly tainted with their thick smell. Two doves burst from a cottonwood and flew low, crossing the evening sun.

The hunters watched the snakes move onto a dam that formed one wall of the tailwater pit. They began to disappear into holes in the dam. That's when Gabriel spotted an enormous diamondback, which he later described as the biggest he'd ever seen.

That remark means something. Gabriel, as it happens, is a lover of rattlesnakes. He used to have a side business catching rattlesnakes and selling them to a restaurant, a whitewashed cinderblock place that usually served burgers and chicken-fried steaks. But in the summer, one night a week, was an all-you-can-eat rattlesnake buffet.

For Gabriel, supplying the restaurant was only a way to make a few dollars out of his hobby. He had already been catching rattlers for years. He eats them, tans their hides, has a quart jar full of rattles. He guesses he's caught a thousand or so. He worked in the oil fields for a while. The vibrations of the drilling rigs drew rattlers in, and they could be found by the dozens coiled on the shady side of a rig. When I visited his home one time, Gabriel opened his freezer to show me a few of his favorite specimens. One was a five-foot diamondback with a pale greenish hide.

You need to know how much Gabriel loves rattle-snakes to understand what happened next. The giant diamondback dashed for a hole. Gabriel grabbed it by the tail. The snake had stuck its head and part of its body into the hole, but Gabriel had a good grip on the tail. The snake was thicker than his thigh, and the part he could still see was substantially longer than his body. He's about six feet tall.

It was a bizarre tug-of-war: the snake trying to slide into the hole, the man trying to hold it out. Gabriel had dropped his rifle to go after the snake. He called to his friend to lend a hand, but the friend only offered a candid appraisal of Gabriel's mental health. Soon the snake outmuscled the man and slipped underground.

With the distraction of the giant gone, the two men noticed something they'd been ignoring for a few minutes. It sounded like a beehive, but louder and deeper. Gabriel had seen a dozen or so rattlers going into the same ground, so he knew there was a den somewhere in the dam, but he'd never heard such a sound before, even from a den. The men looked at the ground beneath their feet and knew it was time to leave.

———

The owner of the ranch where Gabriel and his friend had been hunting wanted to wipe out the den. He had livestock to lose. He poured diesel fuel into the holes in the dam—a common way of killing out a den, though it pollutes the groundwater. Usually, a few gallons of the suffocating fuel makes the snakes come raging out. The rancher had men standing around with shotguns and garden hoes to finish them when they did.

Nothing happened. The rancher poured more and more diesel in through different holes. He kept trying until he'd used 150 gallons. No one could imagine why the diesel wasn't bringing the snakes out. The rancher had killed twenty-seven rattlesnakes on his land that summer, and he suspected a few of his cattle had gone down to rattlers. He wasn't about to give up.

He decided to use a backhoe to open the den. The machine dug six feet before puncturing a cavern the size of a spacious living room. Dark, viscous mud, the kind called "black gumbo" in the oil fields, formed the walls. The floor was a writhing carpet of rattle-snakes, so many no one could count them or even guess their number. Diamondbacks and prairie rattlers crawled across each other's backs. Branching from the large chamber were scores of small tunnels, and snakes moved in and out of these. Panicked by the vibrations and the flood of sunlight, the rattlesnakes set off a chorus of buzzes that drowned the noise of the backhoe. The nauseous smell of agitated rattlers bloomed in the hot air. One man claimed several of the diamondbacks in the pit dwarfed the giant Gabriel had struggled with before.

The rancher brought around a fuel truck with an electric pump. He sprayed a truckload of gasoline onto the mass of snakes, killing thousands. Hired hands burned the carcasses.

This massacre reduced the number of rattlesnakes on the ranch the next year, though a few still turned up. The tailwater pit was polluted beyond use, the run-off draining into the cavernous den. For a long time, in the evenings a strange smell hung low over the cavernful of water where oily rainbows floated. The hired hands who burned the carcasses have all gone away. The rancher and his sons have died, and only an old widow remains on the family land to remember the troubles they had with rattlesnakes.

———

When I was a child my father killed a rattler near our yard. He decapitated it with a hoe, and I watched it pulse for what seemed like hours. The snake, a very small one, kept twitching, even after our white leghorn hens came and started to work at it with their beaks. At dusk I came back (I was forbidden to, but I came). It was still alive enough to shrink from my touch. In the morning, it was stiff. The bubble of blood where its head had been was hard as brick.

Folklore says a decapitated rattlesnake doesn't die

until sundown. The restless one I watched as a child was no aberration; beheaded rattlers often make this lore credible. Their movements don't really stop with the sunset; they diminish gradually, and no moment of death can be specified. But the movements clearly out-last an injury that should, according to everything we think we know, prove instantly fatal. In fact, animal life in general doesn't end as neatly as we expect it to, but this revelation is especially disturbing in the case of the rattlesnake, partly because of the ancestral hatred and fear that cause us to try to kill it. More important, the rattlesnake, unlike other apparently dead animals, is still dangerous.

People and dogs have died from the reflexive bite of a decapitated rattlesnake's head. The venom itself has proven potent even after twenty years of storage; dried, it remains potent for at least fifty. We approach the rattler with such an awareness of its deadly potential that its failure to die neatly becomes terrifying and must be explained, if only in the form of a convenient piece of lore that fixes the time of death.

We would like to think death is a crisp fracture: living, and then not living. In fact, there is no clear division between life and death in any animal. People sit up, fart, and twitch long after they are apparently dead, and an arcane lore of medical and legal specifications has grown up to deal with the practical difficulties of this sloppy division. We debate the merits of machine-assisted life and independent life. When should a person be removed from life support—at the stopping of the heart? at the stopping of the brain?

We ordinary people (that is, not medical professionals) generally believe in the principle of brain death. Brain death as a concept frees us from the responsibility of deciding death, because it is a completely invisible distinction. No layman has the equipment to measure brain activity. With the means of officiating out of our hands, we don't have to decide; we bring our dying to the hospital, a kind of temple, where doctors, in their roles as secular priests, make the pronouncement. They use technology incomprehensible to most of us to make the call, and then they declare a time of death on official documents—as if death happened in an instant. As if it were a crisp fracture from living.

Practical difficulties surround this system. How should we judge cases where the brain still lives, but consciousness and independent breathing and heartbeat are gone? How do we answer the objection that a brain-dead person whose heart still beats can live for many years, as long as someone feeds him? How about the fact that some people who need machine support to live have active minds?

Once a body is declared dead, the difficulties continue. The body may move around. A dead woman may deliver a live baby. Occasionally a "dead" person recovers. In the nineteenth century, one writer estimated that a third of the people buried weren't really dead. He arrived at this figure by considering the number of people who, disinterred for one reason or another, were found to have clawed at their coffin lids or otherwise struggled to escape. Edgar Allan Poe's fictions on this subject were probably more frightening to his contemporaries, who knew of actual cases, than to us well-insulated moderns. The EEG and other advances would reduce the possibility of premature burial, if we hadn't obviated that need by hitting on the idea of embalming our dead and making damn sure they don't get back up. Current embalming techniques involve, among other things, mechanically sucking the juices out of a corpse and replacing them with preservatives. You can also complete a death by burning the body or signing away its organs.

Americans, at least, have taken steps to hide the slipshod workmanship of death completely. We don't sit up with corpses anymore. We give the dying over to hospitals and the declared dead to funeral homes. We don't see our dead sit up anymore; we don't smell them evacuating their bowels like living people. Embalming gives us a corpse two or three days old that looks much like the living person. But that resemblance is an illusion we grant ourselves. It is part of a larger illusion we maintain: that death is a still version of life.

Death isn't still. It is a continuation of what has gone before. The digestive juices in our gut lose their inhibitions and go to work on the organs that hold them: we eat ourselves from within in a last burst of appetite. The bacteria that have been part of our bodies go on living. Suddenly freed to partake of the feast they have always dwelt inside, they prosper as never before. Our tissues, if left alone, take on an array of strange forms as microscopic life converts them; one way or another, the meat we're made of fuels the building of other lives. The blood gels, the breath quiets, the tiny strands of lightning inside the nerve tissues disappear, the form we instinctively like best gives way to other forms; the smells of death and rot are always the smells of small life-forms teeming. Death is real, but it is slow and sloppy; it proceeds in no certain order; its beginning and end are indeterminate; and its causes are not always certain. Dead, we are not stilled; we are activated, changed.

———

On the highway ahead, I see the sinuous curves of a rattlesnake in motion. He moves on the hot asphalt in liquid esses. He is doomed.

Cars and trucks rush by, some of them holding their course, their wheels straddling the snake, others swerving to miss him (perhaps these drivers know the legend of the mechanic killed by a fang embedded in a flat tire). Soon one of them will crush him, by chance or choice. He halts and buzzes on the yellow line. I pull over and wait my opportunity to chase him off the road, but the traffic is heavy. A one-ton pickup finally swerves to hit him.

I watch his body spasm into twisting arcs, the white belly and patterned back showing by turns. It is the old dance of animal flesh: the dying, and the determination not to die.

TARANTULA

Lightning showed them crossing the asphalt in the first tentative cracklings of the rain. They hustled across, more than we could readily count, each brown-and-beige body slung low amid the multiple dark arches of the legs. Rain dappled the windshield. The droplets staggered down through dust and the lambent blood of fireflies. The view at that moment would let you believe you weren't seeing right, that rain and lightning and motion and filthy glass had cooked up an illusion. But the headlights began to show them straight ahead, middle of the road, not just on the shoulders, and the wipers left arcs of clear viewing between arcs of mud, and we knew we were seeing an exodus of tarantulas. We stopped to watch, and as I stepped out of the pickup I saw two of them convulsing on the shoulder behind us, wounded by our tires.

In Africa the wildebeest migrate, and when they come to a river the herd fords. Their great number gives the crocodiles a good chance to feed, but it also guarantees there will be too many for the crocs to take them all. That's the brutal logic of mass movement. This spectacle was a little like that, the tarantulas going somewhere, crossing the asphalt river, and some of them dying beneath the treads of passing cars. One of the cars swerved in a slow ess, as if trying to miss one particular tarantula among the dozens. At the edge of the asphalt we could see them wrestling through the unruly grass to reach the road.

"Males hunting for females?" I said. It was the first time I'd ever seen more than one on the road.

"No, look. They're all going the same way." My friend bent to look at a tarantula breaching the grass at his feet. He was a burly man, but he reached with gentle precision to place a forefinger on the spider's buck-colored carapace, the shell-like cover of the front half of its body. He pressed the spider against the pavement. His thumb and middle finger found a place to grasp in the thicket of legs. He lifted it to show me. It kept its legs bunched against its sides, still, like a cottontail rabbit that thinks you haven't spotted it.

The fangs lay folded. They looked like the parings of

a thumbnail in shape, color, and size. The legs and belly looked soft in their dusty brown hair.

"A female," my friend said. "Look at that one." Another tarantula had emerged from the grass-choked ditch. It walked taller than the one he held; its legs were longer, but its body was smaller and seemed shriveled. As it crossed the puddle of glare from our headlights I saw that its hair was darker, a brown approaching black. These, my friend said, were the characters of a male.

In the truck we found a few empty cups that documented the fast-food places along our route, and a plastic grocery bag. Such trash would serve until we got someplace. We caught eight of them—my friend did, actually. I was afraid to touch them. When we looked at them under the light, we found only two were male, which meant this was no seasonal wandering in search of mates. Only the males wander for that purpose.

"It must be the flood," my friend said. It had rained too much that summer, and the fields were drenched; the place was a disaster area by the President's say-so, the ordinary ditches beside highways supporting populations of duck instead of roadrunner, cattails instead of prairie grass. Some places people were climbing to their roofs or leaving their homes. We stood, our heads bare to the evening's return of rain, watching the tarantulas leave their homes to cross from flooded field to equally flooded field, and then we drove on.

———

Wash your hand and leave it moist. Now place it, tips only, on a tabletop. Let it feel whatever air may move in the room. Do you feel the coolness on your palm? The shiver produced by your own pulse? The tiny shocks made by any other thing moving on the table? Listen with your fingertips, and you'll notice that the tapping of something three feet away comes to you in waves, reaching the nearest fingertip perceptibly sooner than the farthest. If you practiced, honed this kind of perception, you could navigate blind and deaf; you could sense the world by touch.

Now you have a hint of what it's like to be a tarantula. The big spider's entire body is a tactile ear. Some of the hairs are specially built for vibration; rooted in sensory cells, they know the direction of their disturbance. And the tarantula tastes everything, from air to prey, with the tips of its palps and its feet and even with openings on its legs and under its knees.

My grandmother was known as a great screamer. Her screams were startling; they had some peculiar quality of making the front of your backbone itch. She screamed at anything creeping or otherwise verminous. Living in the country, she often happened to see creatures fitting this description. One time when a mouse showed itself in the house, Grandma let loose one of her high-pitched ear-splitters. The mouse ran away at top speed and crashed headfirst into a wall. It fell dead on the spot—of a broken neck, my uncle speculated.

My family claims the scream was responsible for the rodent's suicide. They say the scream was so intense, or so high in pitch, or so something, that the mouse's brain was scrambled. And, to counter my skepticism, they recount The Tarantula Incident.

The tarantula, like the mouse, was an innocent wanderer. It happened into the kitchen where the family were assembled. Grandma spotted it and belted forth. It stopped and lifted its body higher on its legs. She screamed higher and louder; it rose higher. She seemed terrified that the spider didn't run away; it seemed too scared to move. The battle continued to escalate—higher screams, higher spider.

Everyone else was laughing too hard to help. When my uncle finally stopped rolling on the floor, he clapped a jar over the tarantula. This action would normally cause a flurry of legs, but the tarantula simply remained at attention. My uncle soon had the spider sitting securely in the jar on the table. It never moved again. It just stood at the full extension of its legs, and after a day or two someone tossed it out.

Sound is, of course, nothing but vibration.

———

"It's the eyes that gross me out," my wife said. She was explaining why she found tarantulas far more disgust-

ing than black widow spiders. The question came up
when I brought home four tarantulas, my share of the
ones my friend and I had captured on the road. Tracy
has always tolerated my habit of bringing home as-
sorted vermin in jars, but the tarantulas taxed her.

"I thought you would object to the widows, because
they're dangerous," I said. A few dangerous tarantulas
hang out in tropical rain forests, but no American tarantula's bite is dangerous to humans. Of course, getting punctured by fangs that size doesn't feel good.

"I don't love the widows, but they're so shiny they
seem hard. I know they're not, but they look like they
wouldn't be too disgusting to touch." The widow she
had named Sweetie Face sat in a jar on the table in
front of us, snacking on a cockroach. "The tarantulas
have hair, like something you might pet."

I responded with a discourse on the texture of the
widow's hide, during which Tracy left, made something
out of pasta, and returned, still nodding occasionally as
if in agreement with some point I was making. When I
paused she said, "But tarantulas have those two big,
gross eyes. They're so big you can look into their eyes."

"Actually they don't," I pontificated. "They have
eight small eyes, which are grouped into two hairy
patches."

"Which are right on top of its head staring at you."

At that moment the biggest of the tarantulas I'd
brought home seemed to be staring at us with her two

bunches of eyes, from which sprouted tufts of brown hair. Actually, a tarantula's eyes are so weak they're probably only good for noticing the shadows of predatory birds. I had named this big tarantula Prima, after the enormous heavyweight boxing champion Primo Carnera. She stood motionless in a gallon jar. We'd put a cricket into the jar an hour earlier. The cricket had tunneled into the dirt for a while, then come up to look around. It walked under Prima, who lifted herself higher on her legs to let the cricket pass. The cricket made another circuit of the jar and came at Prima from the front, swishing its antennae against her legs.

The tarantula twitched as if she had been shocked electrically, and her twitch raked the cricket to her mouth. She stood holding the cricket the way a dog holds a dead rat. After the first bite, the fangs moved separately, stabbing down and in repeatedly. They were slow, sensual, almost obscene, and the flash of the tips made us shudder.

———

Spiders have pioneered the architectural possibilities for predation. The orb weavers spread nets to the wind. The bolas spider goes fishing with a sticky ball of glue on a string. Some orb weavers spread sheets of flypaper and then attack snared prey from below. One spider spins a net and then, camouflaged as a plant stem, sits with the net stretched between its front legs, waiting to cast it on an insect. Some orb weavers make their snares three-dimensional by pulling on the hub until the web deforms into a cone: when a flying insect hits, the spider releases the hub and the web snaps back, wrapping the insect by spring action.

The range of snares is astounding, but the spider clan knows other fancy predatory tricks too. Some spiders have bodies crusted with nodules and projections, their graceless forms disguising them from their prey and their predators. Others wait within flowers that match their own colors, killing the eaters of pollen and nectar who fail to see them. Still others trap prey from a distance by spitting toxic glue.

All of which points out how simple the tarantula is. While other spiders have evolved complex predatory behaviors, the tarantula still earns his meals the old-fashioned way: he hides in his hole until he senses some hapless critter passing, leaps on it like a mugger, and mashes the hell out of it with a wicked set of mandibles. He is to the orb-weaving spider as the *Australopithecus* is to modern man. He doesn't know any tricks involving color. He doesn't build snares. He hasn't even mastered the fine art of killing with a dainty, pinching bite full of venom and then sipping his victim down like a gentleman. He has venom all right, but he still chews his food like a wad of tobacco, slobbering digestive juice all over it as he goes.

It's not that the tarantula lacks silk. He uses the stuff

to line burrows. He lays a few trip lines on the ground, radiating from his burrow starburst-style to tell him when something's walking past. The female knits a silk knapsack for her clutch of eggs. The tarantula's use of silk would actually be impressive if his younger cousins weren't such geniuses with it.

———

As I write, the world is crazy for dinosaurs. Especially the carnosaurs, like *Tyrannosaurus rex*. While paleontologists argue whether the carnosaurs actually were predators—some say T. rex must have been a scavenger—Hollywood and its affiliated toy manufacturers continue to hype them as the baddest beasties ever to walk the planet. I'm sure it's good marketing.

The movie people don't mention the animals that ate carnosaurs.

In Georgia and Alabama a paleontologist found only a few bones of a big carnosaur called *Albertosaurus,* which was common elsewhere in that period. The Albertosaurs seem to have been edged out of this swampy environment by crocodiles. Bones of young Albertosaurs show the hack marks of crocodile teeth. This evidence has been lying around since the time when dinosaurs "ruled the world." The type of crocodile that ate Albertosaurs grew to about thirty feet; larger types elsewhere managed fifty feet. They were longer, and probably more massive, than the Tyrannosaurs.

The biggest modern crocodilians don't grow over thirty feet, but that's still bigger than most of the dinosaurs. The crocodile eats whatever it wants, from fish to porcupines. The African crocodile, which is not the biggest variety, swallows warthogs whole. One was seen pulling a two-ton rhinoceros into the water by its snout before drowning and eating it. In World War II, Allied forces trapped a thousand Japanese infantrymen in a stretch of mangrove swamp on an Indonesian island. Twenty of the thousand came out alive, most of the others eaten by crocodiles in the night. Crocodiles still reduce the world's human population by several thousand each year. They take people in Africa, India, Australia, and Southeast Asia. Once in a while even the supposedly harmless American alligator eats somebody.

The crocodilians have been around for 170 million years or so. From the time of the dinosaurs to the present "rule" of humanity, the crocodile has been swimming quietly in rivers and oceans, dining on members of the ruling parties, spilling more royal blood than Richard III. The reptile clan have specialized in many different ways, even introducing a successful line of venomous snakes—which the crocodilians eat, along with such venerable reptiles as the shell-protected turtle.

The crocodile's success debunks a common misconception about evolution: that further-evolved animals replace outmoded, unsuccessful ones. Even though reptiles (and mammals and birds) have evolved an enormous number of successful body designs since the croc-odile came on the scene, the crocodile's "primitive" design has never failed. People like to think evolution means getting better, when all it really means is getting different. Extinction is not bound up with evolution as tightly as one might think. In fact, except for a few special cases, we don't know why species die out. It's a hot topic in biology.

The "primitive" predators likely to flourish un-changed while their descendants evolve into far differ-ent forms usually have a simple design, which usually means a big, dangerous mouth. Such predators are gen-eralists, capable of killing a wide variety of prey. Con-sider the crocodile and the great white shark, both mil-lions of years old without substantial change. They don't kill with snares, cooperative group hunting, or even much intelligence. They just bite, often killing by sheer mechanical injury.

Experts call the tarantula a primitive spider not only because of the skills he lacks, but also because of a few anatomical differences from the later models. (He's not the most primitive spider—a big, tarantulalike Asian critter, with a segmented abdomen, holds that honor—but the tarantula is the most primitive spider you're likely to see.) Like the white shark and the crocodile, the tarantula is a simple generalist with big fangs, and a long-term survivor. Today it prospers around the world, in habitats from desert to rain forest.

The tarantula's wait-and-ambush lifestyle has proved so successful that the spider clan has since rediscovered it many times. For example, the entire group called wolf spiders are members of the later, "advanced" type. They're not related to tarantulas; they're web-makers who went back to living in burrows and springing on passing prey.

Late one night, as I sat up reading, a spider the size of my palm sauntered out from behind the couch. It was a rabid wolf spider ("rabid" is part of its common name). After a chase and much dishevelment of the furniture, the rabid wolf became a guest in my water jug. I fed her for a few days. She took prey the way a tarantula does, pouncing and seizing. Eventually I introduced her to Prima, her even bigger neighbor in the next container. The rabid wolf didn't last a full second.

I thought of the ten-foot crocodile whose stomach was found to contain a four-foot crocodile. For the simple predators, size is everything.

———

I caught June beetles for the medium-sized female we called Harriet. The first beetle I put into her cage waddled directly to her as if offering himself for sacrifice. She put a foot on his back, tasting him at a distance to see if he was worth killing. When she seized him he broke with a sound like a walnut cracking. I tossed in another June beetle. Without dropping the first, she snatched up the second, crunching both into a ball, which she continued to work with her fangs.

I put in another beetle. And another. She kept snatching them up, never dropping a morsel. I had caught only six, and soon the whole half-dozen hung from her fangs in a ball glistening with digestive fluid. Her fangs worked away, more like machetes than the hypodermic needles spider fangs are usually compared to.

Food starts to fall apart after a few seconds of such treatment. Harriet's wad of June beetles was apparently becoming too sloppy for her taste. She put it on the cage floor and stood over it, arching her back so that her spinnerets aimed straight at the mess. Those two spinning organs on her hind end worked with the dexterity of human fingers as she threw fine silk. She rotated in place as she spun, her spinnerets always aimed at the mess of beetles. Soon she had webbed the mess into a neat bundle. She settled down to suck the juices out.

———

The tarantula is the largest and strongest spider. The largest and strongest kind of wasp is called the *tarantula hawk*. The feud between these two giants fascinates people who follow wildlife; it is among the naturally dramatic predator rivalries, like cougar and coyote, lion and hyena, sperm whale and giant squid. The spider is primitive, a generalist predator whose kin have long since evolved tool-using specialties and left its brutal hunting methods behind. The wasp, far from primitive, is one of the most specialized animals on earth, and its specialty is hunting tarantulas. Their encounters usually end in the most horrible death imaginable.

As I write, the particular tarantula hawk I want to tell about sits on my desk, dead. I captured it in a gravel parking lot outside a truck stop a few years after the rainy summer during which I kept my first four tarantulas. Measured in a straight line (rather than along the many curves of its body), it is a shade under two and a half inches long from the outer coils of its antennae to the tip of its stinger, though the stinger of this one is mostly retracted. When a hawk is planning to sting, its stinger protrudes another quarter of an inch. This one's veined wings span three and a quarter inches—bigger than the rim of the coffee mug sitting beside it. Its legs are longer than its wings. In warmer, wetter places some hawks double these dimensions.

———

Trucks had parked on this lot for years, and the previous night's rain had summoned the smell of petroleum from the ground. The hawk flew in figure eights over the landscape of gravel, mud ruts, oil stains, and sickly weeds. As she skimmed near me, I caught a whiff of an odd smell like that made by ants when they sting in a mass. Touching down abruptly, the hawk walked in rapid zigzags, her coiled antennae wiggling. Her black body seemed surrounded by an orange cloud, as if electric shocks were exploding around her so fast as to be only subliminally visible. Only when she stood still could I see this aura had been made by the motion of fast-beating wings. Their dull orange color was one I'd seen elsewhere only in wood fires, at the hazed border of flame and smoke.

Her zigzag walk reminded me of a bloodhound. Abruptly she took the air and circled, her circles lowering and tightening toward the spot she'd taken off from. She landed and walked a circle of a few inches, then vanished.

I came closer. The hole in the ground peeked from a camouflage of gravel and withered weed stem. Barely perceptible traces of silk tapered from the hole. The hawk came backing out; a tarantula came out facing her in a lockstep dance, his impossibly large shape seeming to unfold from the slender opening.

The hawk led the dance away from the burrow. The tarantula followed in a dreamy slow motion. The hawk stopped backing and began to feel the spider all over, like a shopper testing the produce. She crawled the tarantula's body, even turning upside down to scoot underneath like a mechanic checking for leaks. The tarantula obligingly lifted himself higher to give her room.

Suddenly the tarantula shrugged off his hypnotic stupor and made that fatal twitch, snatching at the hawk, but the hawk had already positioned herself to advantage. The two rolled over and over, knocking bits of gravel around. Suddenly the tarantula contracted and flopped onto his back. He had been stung, probably in the juncture of a leg with the cephalothorax. His legs flexed rhythmically, as if washed by invisible waves.

The hawk flew off toward a fringe of weeds that grew along a chain-link fence. I came closer and prodded the tarantula with a weed stem. He kept flexing, unaffected by the poking. I saw the hawk rise from the weeds and backed off to a respectful distance. I have heard the hawk's sting is incredibly painful, far worse than the fiery stab of a paper wasp.

The hawk returned to the tarantula. Taking a massive leg in her jaws, she lugged the body toward the weeds. When she had gone less than a foot, she once again abandoned the body to visit the weeds. I followed her, but couldn't see exactly where she went. I had seen other kinds of predatory wasps tackle smaller spiders, and in one species I had noticed the odd habit of checking the prepared grave, a deep tunnel, repeatedly. The wasp sticks her head into the tunnel and flitters her wings. Maybe she's making sure no prowler waits to catch her when she is burdened with groceries, or maybe she's just resting.

I guessed the hawk suffered from the same compulsion. She dragged the tarantula a little farther on each trip, frequently pausing to visit her hidden tunnel in the weeds. I knew what fate awaited the tarantula there. The hawk would deposit an egg on his abdomen. A wormlike larva would hatch from that egg and devour the tarantula, taking a few weeks to finish the huge meal and saving the major organs for last.

Of course, a dead spider would rot before the larva could hatch. That's why a mother hawk carefully places her sting to leave the tarantula alive but paralyzed. Spiders removed from wasp burrows have been found to last at least nine months in this condition. In the burrow, however, the larva soon eats some major organ and kills its host. Grown fat and indolent, the wormlike parasite falls into the sleep of pupation. The spider's hairs and carapace decay in the soil as the larva at their center transforms into a gleaming hunter.

It was too horrible to allow, so I decided to intervene. Either that, or I wanted the two combatants for my collection—who knows his own motives? Anyway, I rummaged in a nearby garbage barrel until I found a plastic margarine bowl with a lid. When I returned, the hawk was lugging away at the tarantula noiseless and patient in his paralysis. I clapped the bowl over the odd couple. The hawk buzzed and thumped against the plastic. Somehow I worked the lid onto the bowl, capturing the hawk and leaving the tarantula out.

The hawk died in a day or so—they usually do in captivity. Now her cat-eyed, hunchbacked, long-legged carcass sits on my desk. Once in a while I stare at the stinger and contemplate my stupidity.

The tarantula's fate was far stranger.

I brought him home and laid him out on a paper plate. He was only paralyzed, so reviving him should have been possible, I reasoned. I moistened the bristles around his mouth. I looked him over with a magnifying glass, trying to find the stung spot, but I never did. I poked him with a pencil. I blew on his belly, trying to push oxygen into his simple lungs. That was all I could think of to do, so I left him alone.

I decided I might as well consider him dead. I planned to preserve the tarantula, but I didn't get around to it for a while. The matter was not hygienically urgent because, unlike a truly dead animal, this one wouldn't rot for months. When I finally went to stick the tarantula into a jar of vinegar, I found him standing up on the paper plate cleaning a forefoot with his mouth. As soon as I felt sure I wasn't really having a heart attack, I prodded him into a cigar box. He lived in a terrarium for about a month, walking about in a tentative, uncoordinated gait, occasionally bursting into a spastic frenzy of gestures. He ate an occasional grub, but soon began to shrivel. I found him lying flat one morning, soft as a silk glove.

———

If you put a predatory wasp, even a tarantula hawk, into a tarantula's cage, the tarantula makes a nice meal of it. A tarantula could do the same to a hawk that attacks it in the wild. Those ragged scimitar fangs would reduce the wasp to pulp in a second.

The hawk is fast, and it's one of the strongest insects. Its long legs, movable head, large eyes, and shearing mandibles give it abilities reminiscent of that awesome predator, the mantid. But it shouldn't be any match for the tarantula, which can be ten times heavier. The hawk has its sting, but the tarantula has its venomous bite: either one can paralyze the other with one shot.

Somehow the hawk casts a hypnotic glamour over the tarantula. So strong is this spell that some species of hawk dig their burrows only after finding and examining the tarantula, which means the tarantula simply stands around for an hour or so waiting to be stung. The hawk only casts its spell when it's specifically hunting tarantulas; that's why it gets eaten if you capture it and throw it to a captive spider. Furthermore, each species of hawk seems to match a species of tarantula; a hawk can't hypnotize the "wrong" kind of tarantula, and if it tries, it dies—the ultimate in predatory specialization.

Hypotheses to explain the spider's trance are numerous. Maybe the hawk's odor reveals that it's emitting a hypnotic chemical. Maybe the hawk's pattern of touching the spider hypnotizes it—a method used by Franz

Mesmer, the pioneer hypnotist of human subjects. Maybe the tarantula is a simpleton, responding to stimuli in a few stereotyped ways, and the hawk knows how not to trip its fight response. Maybe the hawk mimics the sexual advances of another tarantula.

We really don't understand how the tarantula hawk entrances the tarantula. But then, we don't understand how human hypnosis works, either.

———

"It was nailed to a tree out on my granddaddy's farm," my friend said. "I suspect Satanists. Anyway, I thought you might want it. Nobody else would."

A dog's skull. On top of the domed region that used to encase a brain, the round entrance wound of a bullet. On the hard palate beneath, the exit wound, smaller than a nickel. The dog must have been shot point-blank. Maybe it had been rabid. Now the skull was bare, picked and long since abandoned by ants and carrion beetles. Its long cuspids wiggled in their sockets at my touch.

I knew what to do with it. I was making a terrarium for Harriet. The sight of her crawling over the dog skull proved scintillating, though the dollhouse furniture a niece donated to the cause created an even eerier effect.

On Harriet's first day in the terrarium, one of my friends reached in to pick her up. She didn't like the idea. She bent herself up at an oblique angle to show her fangs and the light hair around them. When my friend persisted, she ran around the terrarium, kicking hairs off her abdomen with her hind legs to discourage her tormentor. Soon after he stopped—without having held the tarantula—his forearm broke out in itchy little nodules. I hadn't been able to see the hairs as she shed them. I've since seen pictures of tarantula hairs magnified; each filament looks like a cross between a harpoon and a Christmas tree, and they work their way into the skin, causing horrible discomfort in some people, no effect at all in others. Later, staring at the bald pinkish patch on Harriet's abdomen, Tracy said, "I take back what I said about the eyes being the grossest part."

After Harriet had lived in the terrarium for a few days, I sat watching her where she snuggled in a hollow place in the dirt beneath the dog skull. "I wonder if they're really territorial," I said to no one in particular. The next thing I knew I had embarked on another of my unscientific experiments, dumping in another tarantula, a stubby-legged, slightly immature female we had captured on the road that rainy night. They both reared up and tangled their legs, fangs stretching open wide enough to accommodate a human thumb. They stumbled over the dog skull in their grappling. They unclinched and scrambled away to opposite corners of the terrarium, knocking doll tables and chairs over in their haste. In the wild, that would surely have been the end of the fight, but they were cramped together in a cage. I thought of reaching in to take one out, but both were riled, and I was afraid to try.

They met again, seizing, pushing, threatening with fangs. The spectacle was like amateur wrestling with meat cleavers. They separated and retreated. Each felt her way around the terrarium until they met again for a similar skirmish. They didn't seem to grasp that they were caged; they repeatedly ran away in a panic, only to bump into each other and brawl again. I stayed up well into the night listening to music and watching them fight.

In the morning the stubby-legged female lay dead, her tough carapace bitten through like piecrust. Harriet didn't seem to have eaten on her: it was a simple killing.

"The females are highly territorial," I told Tracy.

———

Snooping under rocks down by the creek, I turned up a meager supply of tarantula food: one small centipede and a dozen pill bugs. I doubted a tarantula would eat such small prey. I tossed the whole batch to Prima. With a precision I didn't expect, she stabbed her long fangs into the skinny centipede. After she had worked at him awhile, he looked like a bit of knotted thread. She had no interest in the pill bugs; when one bumped into her, she shook her leg like a cat that has stepped into a puddle.

It's common for a tarantula to pass an entire winter without a morsel, but I didn't know that. I thought she needed food. I went to a pet store and bought a dozen crickets.

I poured the entire dozen in at once. Prima stood in the center of the jar. She immediately took a cricket, folding his little brown-and-white body double with the force of her bite. His comrades didn't take any particular interest in his fate. They spread out around the jar, hugging the glass. The pet store must have skimped on the protein: each cricket seized a pill bug, turned it upside down, and began devouring it, using its shell as a bowl and eating the insides. The pill bug (also known as a wood louse or a roly-poly) is a crustacean; maybe crickets prize them as much as we do lobster.

Prima stood in the center of the ring of carnivorous crickets, chewing her cricket slowly. She looked like the queen of some little pocket of hell.

———

The fourth tarantula I'd brought home that rainy night was the only male. We called him Raoul. His behavior wasn't like the females'. If I prodded Harriet with a pencil, she would move off a few leisurely paces. If I prodded Raoul, he would run six frenzied laps around his container.

One day Raoul spun out a rectangular sheet of web. It was about the size of his own body. He lay on it with his belly pressed to the silk, bouncing and trembling. Then he rose and applied his palps to the droplets that had oozed from the pore in the forward part of his abdomen. His palps swelled at the tips. The male tarantula uses the palps, which aren't connected to the gonads, to copulate. Until humans invented artificial insemination and in vitro fertilization, only two animals used such a two-stage sexual system—the spider and an insect called the mantidfly.

Raoul circled his container slowly, as if afraid to bump into anything, his gait different from before, eager but somehow delicate. It was such a change from his former hyperactivity that I felt vaguely embarrassed to watch him.

A female tarantula takes a decade or so to mature; then she can mate every year for the rest of her life. She can live to the age of twenty or thirty. A male spends his decade growing up, gets a few days in autumn to find a mate, and then, successful or not, he dies. I decided to give Raoul the same chance he'd have in the wild. I put him into Harriet's terrarium.

They met at what used to be the dog's nose. Harriet flinched and took a step back. ("Egad! A tarantula!" Tracy said, guessing at Harriet's first impression of her suitor.) In the wild, a male has to stomp around outside a female's den acting like prey so she'll come out. Here, however, Raoul could proceed with the tapping of Harriet's forelegs. The tapping forms a code, particular to a species, which seems to translate as "Please don't eat me." It is not terribly uncommon for the female to ignore this request.

Harriet reared up on her hind legs to expose her fangs. Raoul reared up to meet her. They wrestled, just as Harriet and the stubby-legged female had a few days before. The difference this time was Raoul's thumbs.

The male tarantula's thumbs appear with his final molt, the one that makes him an adult. They project from the knees of his forelegs. I had hardly noticed these little spurs on Raoul, but now they came into play. He neatly hooked one thumb into each of Harriet's fangs, so that his long front legs leveraged the fangs apart. He was safe for the moment. He pushed her back until she was almost vertical; then he plugged one of his clubby palps into her belly. They stood still in this position for several minutes. Then he pulled away and exploded into a flurry of legs, bouncing off the ground repeatedly in his hurry to escape. Gone was the awkward gait of his mate-hunting phase; returned was his hyperactive paranoia. He ran straight up the glass wall and cowered in a high corner. Harriet strolled off, looking a little irritable, but she showed no further interest in Raoul. This fact did not persuade him to descend. He stayed there all day and all night, pressing himself into the smallest possible volume.

Raoul had mated and would soon die. Prima, languishing in captivity, would have no opportunity to mate. Harriet would soon produce an egg sac, and then the problem of caring for many tiny tarantulas would descend on me. Tracy and I were about to move to a new city, and I decided the best move would be to free the tarantulas. I released them in a grassy field. I like to think that field is pocked with the burrows of their descendants.

Every few years I get the tarantula itch. It's hard to find the females, but the males are easier. All I have to do is pay attention as I drive highways in the summer. I notice where I see a tarantula. Then, in the last few autumn days before the first freeze (which tarantulas predict more accurately than human forecasters do), I visit these spots. The males who haven't mated yet are desperate by now. They cross the highways in search of females. Sometimes the road is dappled with their crushed bodies.

I cruise these spots at sundown, watching for the eerie gait of the tarantula. If I can stop safely, I get out and stand on the shoulder to meet the wanderers. I usually catch them in baby-formula cans, shooing them in with the lids. These males, who would die in the freeze, can live a few weeks longer indoors. As I drive home, I can hear the spiders scratching their tin prisons on the seat next to me.

———

Acquaintances of mine who often drive Highway 160 call the part east of Meade, Kansas, "the Stretch," because it is a desolate, boring road. The towns are small and thirty miles apart. There are a few places where water has gouged wicked scars on the red earth, but otherwise the view is all field, fence, cattle, open sky, and asphalt, occasionally interrupted by a steep hill or a tight turn. To drive that road is to stretch your patience, or your ability to stay alert. Everyone speeds through.

Driving from the east, one couple was nearing the end of the Stretch. As they came to a bridge near Meade, they saw something that cured their boredom.

"It must have been this big," the woman said, holding her hands in a circle eight inches in diameter. She claimed the tarantula was far bigger than any other she'd ever seen, so big it made her nauseous. She felt conflicting urges to ask her husband to swerve and kill it, or make sure the car didn't touch it. When they had passed she saw the tarantula in the side mirror. It was striding off the road just beside the bridge.

She said she had no love of spiders, but had never had such a visceral reaction to one before, even to another tarantula. "It was the size," she emphasized. Her husband nodded vigorously as if to underscore her remark.

Of course, I doubt the monster exists. Maybe weari-

ness and surprise enlarged the spider in my informants'
perceptions. But it's possible. Something about the psy-
chology of the woman's reaction strikes me as authentic.
That primal arachnophobia most of us have a bit of
emerges in direct proportion to the size of the spider.

I covet that tarantula. Others have told me their stories of tarantulas on the Stretch. The creatures, they say,
are plentiful and large there. I can vouch for the abundance. I have spent autumn evenings cruising that road,
my eyes scanning the asphalt for the distinctive rippling
movement that is the walk of the tarantula, a movement only slightly different from that of a maple leaf
pushed along by wind. I have seen tarantulas on the
Stretch, and, on rare occasions when there's no traffic
on the shoulderless road, I've stopped the car to capture
one, typically a male with a leg span of about five
inches.

But I've never seen the huge tarantula. It would have
to be a female to get that big, so maybe it's still alive
these several years later. Sometimes I park near the
bridge and walk along the creek it spans, gazing down
the steep bank into the dark deepening with evening,
and I hope for a monster.

P I G

A hot place, the abundant foliage moist and gleaming like emeralds in the morning sun; deep humid shadows; at a little distance in any direction, visible wisps of steam navigating sinuously among the fronds of palm and fern. Blunt-winged butterflies the color of ripe cantaloupe flesh fumble drunkenly at massive flowers. Something larger is moving: a bird picks its way over the mossy ground, its massy hindquarters bobbing with its gait. It is like no bird you have seen, a sort of drab, bloated dove with useless, stunted wings and a buzzard's hooked beak. Its frame is thick-muscled, its walk odd but hardly the clumsy stumble we have always read about.

Out of the undergrowth lunges a bigger animal, four-footed, dark-furred, with a high, ridged back and a long snout not unlike a rat's. But this is no rodent. When it has seized the bird in its mouth, its forward momentum stops, and you can see that it's a pig. Perhaps you're surprised by its lean build; perhaps the snout seems unnaturally long—it tapers in and then flares out again before ending in a flat, cartilaginous disk of a nose. At the moment, however, what you notice most is the set of tusks curving out around the snout, which the pig, with short twitches of the head, is using to gut the panicking bird. The small wings beat futilely; the bird's bright black eyes dart around and then become still. The pig is already headed into the sheltering green with his meal.

This was a typical death for a dodo. A fast runner, the bird was crippled by a trait common among animals that evolve on islands without predators: it had no fear of strangers. It died as hundreds of other island animals have, hunted down by predators whose standard of brutality was beyond their experience.

The dodo, a relative of the dove, died out about three hundred years ago—it's hard to be exact because no one noticed it at the time. The reason for its extinction is complicated. Most people think human predation killed off the dodo, but that's only part of the story. The dodo actually succumbed to competition and predation by a half-dozen or so invaders of its isolated habitat. Though none of these animals springs to mind when you hear the word *predator,* together they precipitated a dodo apocalypse. The culprits, which all arrived on ships, were the goat, cat, rat, monkey, human, dog, and pig.

Goats gobbled up the fallen fruit dodos liked to eat; they were competitors. The rest hurt the birds more directly. Contrary to legend, the dodos were graceful runners and could deliver a painful bite, and there's one recorded instance of a man venturing near a nesting site and receiving a sound pecking from a gang of dodos. But these fighting and running skills didn't go far enough to protect the dodo's eggs and young from sneaking predators like cats and rats. In his recent book *The Song of the Dodo,* David Quammen presents good evidence the macaque monkeys that had already invaded Mauritius three hundred years ago are even today pressuring a certain endangered kestrel species by eating its eggs. The kestrel makes its nest in cliffs; the dodo's nest, lying on open ground, would have been an easier target for the macaques. (Don't ask why somebody brought monkeys to Mauritius; nobody knows.)

Humans shot the big birds or simply ran them down like barnyard chickens. Their bodies were so packed with edible meat that crews could provision their ships by gathering dodos for a few hours. Dogs, like rats, ate the eggs, which were bigger than those of geese. The dogs delighted in killing the adults as well.

The pigs on the islands were introduced in the hopes that they would establish a self-supporting wild population. Then they could be hunted by settlers and the crews of passing ships. The pigs roamed free, grazing and rooting for food. They were captured only for slaughter. The wild population took hold as planned. This success is hardly shocking, since the "European" wild boar has established branch offices everywhere from the East Indies to West Virginia.

A wild pig is quite a different animal than a captive one, and the captive pigs of today are much different than any, wild or captive, that lived three hundred years ago. A wild pig, even when well fed, looks lean and razorbacked. It's long-snouted and hirsute. It rides high on thin legs. It has a set of tusks perfect for eviscerating other animals. Domestic pigs have tusks too, but most farmers trim them for safety reasons.

The tusks can amputate a human limb. Once, a friend of mine helped two friends trying to restrain a huge domestic boar so that its tusks could be cut. To control a pig, you put a loop of wire around its sensitive rooter (the flat disk at the end of its snout) and pull up. The pig holds still for fear of hurting himself. This boar caught on to the plan and tossed the men aside, refusing to let his nose be wired. One of the men decided to knock the boar unconscious so they could get on with the clipping. He swung a two-by-four at the boar's head. The boar bit it in half in midswing. The men apologized and left the pen.

Pigs are omnivorous, living on nuts, roots, fruit, grain, bark, and carrion. They even eat chunks of coal, crunching it like hard candy. They also prey on live animals. Their menu of common prey items includes insects, frogs, lizards, snakes (even rattlesnakes), rodents, and lambs, but they will try almost anything, including fish and crabs and large land animals. In Argentina, where the boars get especially large, they prey on domestic rams.

Pigs were probably the most damaging predators for the dodo population. They were big and fast enough to hunt down the adults, but they would also have taken an omnivorous interest in the eggs and young. They had keener noses than humans and could hunt dodos even in the jungled parts of the island. Unlike dogs, they didn't stay near human habitations; they avoided humans. They could strike anywhere. For a flightless bird that had never seen predators before, the pig was catastrophic.

The dodo holds a peculiar place in Western consciousness. It was the first animal we realized had become extinct in historical times. We understand it as an emblem of extinction, but few of us know anything else about it. For example, most Americans think the dodo was Australian, though it actually lived on the island of Mauritius.

The dodo is our emblem of extinction, but the myths that have grown around its demise are revealing. We have painted the dodo as a clumsy, freakish creature that deserved to die out—a myth so important to our idea of evolution-as-competition that it's taught as fact to schoolchildren, so pervasive the bird's name is a synonym for *fool*. It's as if we must find deserving losers in the competition. This myth is probably a salve for our collective guilt about the way we treat the world around us, destroying other living things through our clumsy greed.

Paradoxically, the competing explanation, in which humans bear full responsibility for the extinction, is yet another bit of arrogance. It blames us for the actions of the so-called domestic animals that also exterminate species. In Mauritius, we can hold ourselves responsible for bringing in alien animals. But pigs, dogs, and cats are always exploring new territories, to the detriment of native species. The process is a necessary corollary of natural selection. We are mere accomplices.

We arrogantly see nonhuman animals as innocents at play in nature's temple, potential victims of evil invading humans. But cats kill for fun, wolves slaughter more than they can eat, and pigs destroy the vegetation they depend on. Many animals are just as intemperate and greedy as we are, though we accomplish more in the way of destruction.

The facts force us to set aside ideas like "master," "pet," and "ownership" in favor of a subtler model. We do not "own" domestic animals, except in a legal sense that's meaningless to other animals. We are their partners in an odd symbiosis that has altered us all—hu-

man, canine, porcine, feline, and the rest—into an ag-
glomeration of strange species, a radical system that has
remade every ecosystem it's touched. If we have domesticated the other mammals, they also have domesticated
us.

————

The chores done, we sat on the fence watching the castrated pigs at the trough. The weather had been dry
and the dirt was in a fine powder when the rain began
to fall. The drops were large. Each one sent up a puff of
powdery dust. The pigs didn't seem to mind the warm
rain, and neither did we. After a while we went to a far
corner of the pen to look at the spot where Cecil had
thrown in the skunk. He had found the skunk that
morning, road-killed the night before. All three of us
had watched the pigs eat it, but we were still incredulous. Some black hair still lay on the spot, but that was
all. No flesh remained, nor any bone, nor even the almost ineradicable smell of the skunk.

We decided to ride the cut pigs. Cecil went first. The
pig he mounted turned a one-eighty and ran screaming
under the sheet metal shelter, neatly scraping Cecil off.
He rose covered with dust, a few fat raindrops giving
him polka dots. "Very smart," he said. "The horse
never thinks to do that." There were a few further
attempts at pig-riding, all of them brief.

We went to look at the sow. She lay placid now, eight

piglets crawling over each other for access to her dugs. Earlier we had seen her run to the trough, her quarter-ton bulk thundering with graceless speed, the piglets dropping from her dugs to squeal and writhe in the dust. No one suggested riding her.

"There were nine piglets," Jim said. "She ate one."

"He was a runt," Cecil said, as if defending her. We watched. A sound like a distant waterfall rumbled in her throat. Her ears lay forward, concealing her eyes. On the back of one ear I saw half a dozen dark bits of shrapnel—fleas. I could see her skin through her sparse russet hair.

"That one old man fell down in his sow's pen, and she ate him," Cecil said. "Somebody went to check on him of an evening and found him. Found his bloody clothes, anyway."

We watched the piglets paw at the sagging dugs.

———

Human predation on wild pigs goes back at least as far as our cousins the Neanderthals. Pigs are tough to hunt because their noses and ears are better than ours, and because they're smart. Cornered, they fight viciously. Ancient falconers made a safe sport of the pig. They sent golden eagles to attack from above, while they watched from a distance. There was no chance of the eagle bringing down the boar. It was pure blood sport.

The European sport of hunting wild boar spread

with the colonial expansion of Europeans and pigs. Massive dogs bred for silence and strong jaws follow the boar by scent. Mongrel dogs follow the boar dogs. Rude, cowardly, uncivil, these mongrels will sound the cry when the boar dogs have brought the quarry to bay. Men on horseback follow the mongrels.

The boar: crescent-moon tusks, uppers honing the lowers with each hostile gnash of the jaws. The massive boar dogs look like ants in a big boar's fur. They clamp onto ears and snout and are tossed by the tossing head. The tusks seek any unprotected flank. They aren't horns, they don't stab in; they bite. Some dogs will die before the human hunters catch up. The wounds are long and clean, a desert highway drawn in blood.

The men hear the yelping retreat of the mongrels. A hybrid dog is truer to itself than a purebred. It values its own life. No breed like the gigantic boar dogs could live long without human intervention. They are bred for suicide.

Arriving at the scene, the men see what they can see—perhaps nothing but dogs dying. If the dogs have the boar cornered, the men must face a charge. In old Europe they carried a long pike to meet the charge. It was equipped with a crossbar to stop a stabbed boar from coming all the way up, because pain certainly wouldn't.

A man on a horse was not safe. The charge of a big boar, say four hundred pounds, would flatten a horse. A near miss might be close enough for the boar to scissor the horse's belly open, literally spilling the guts. A boar screams like—well, like a stuck pig. Like the shrill of a human child, combined with a cavernous rumble.

So much for the safest way to hunt boar in medieval Europe. There were other methods for the brave, including hacking at the boar with a sword from horseback, if you could afford to lose your horse, and your life.

———

Pure hunting of swine has never vanished, but it gave way to a subtler relation, a sort of partnership with one party less than willing.

The partnership of human and pig must have started when we settled down. We stopped following the migrating herds and stayed in one place, somewhere with firewood and running water close by, and when we did, our fatal flaw began to show, the flaw that best defines us to this day: our habit of making trash. Old bones and gristle, unusable scraps of hide, the feces of human and dog (the dog was already with us), broken tools of wood or stone or bone, plants cut to make room for our shelters, the husks and shells of our food—we made middens. Smorgasbords for the scavenger.

Pigs came for our refuse, rooted out our corpses, raided our crops. We had set an accidental trap: everything we cast off or set around us drew pigs in, and when they came close in the night we could kill them. They were good eating. Almost every part of a pig is edible, from feet and tail to that gristly disk of a nose, though some parts have to be soaked or simmered longer than others.

Conversely, the pig finds us utterly edible. Theories about the religious origins of burial tend to overlook the emotional realities people faced. We didn't want to witness the eating of our kin: that's why we began to bury them.

The folk wisdom of the late twentieth century explains the Jewish and Moslem taboos against pork as a recognition of parasites in the meat, but a better explanation lies in the stony ground of Palestine. There, people could not always bury corpses simply by digging down through topsoil. Cavities gouged from layers of limestone exposed in cliffs; natural caves; hollows beneath the roots of great trees—these held the dead. The tombs were sealed with rocks. A family owned its tomb for generations, so that when the Old Testament says a man went "to lie with his ancestors," the image is more literal than may be readily apparent. This is the land where pigs were called unclean, unfit to eat or even to touch, because in this place the pig, that rooter of human leavings, devoured corpses, and the problem of keeping swine from the dead was never entirely solved.

Other cloven-hoofed animals, like the cow, were considered edible, but the pig wasn't, because it "cheweth not the cud": meaning that it does not have multiple stomachs, meaning that it doesn't restrict its diet to vegetable matter, meaning that to eat the pig is to eat the human flesh it may have eaten. The taboo colors our image of the pig. William Golding made a pig Lord of the Flies (the term itself is a biblical demon-god's name) in his novel of human degeneration. You see the taboo evoked in horror fiction now and again, the pig as eater of human flesh, or the human mistaken for pig and eaten by his friends.

In the book of Isaiah, the sins of a heathen people include sitting "among graves" and eating "detestable things"—mice, the blood of "unclean" animals, and the flesh of swine. The link between pigs and graveyards comes up again in a peculiar New Testament story of exorcism. Arriving by boat in a country abutting the Sea of Galilee, Jesus meets a man possessed by demons. Three different writers relate this story, and the details about this man vary according to the teller of the tale. He is variously naked, given to self-mutilation, a howler in the night, and a breaker of the chains his countrymen tried to bind him with. No doubt in later times he would have been called a feral man, and in the twentieth century a schizophrenic or an autistic. Matthew claims there were two such men, whom he identifies as swineherds.

All the writers agree the possessed come to Jesus from the tombs of the dead, where he lives as an outcast. He describes Jesus' attempts at exorcism as torment, and names himself "Legion" because many demons inhabit him. Striking a bargain with the demons, who are reluctant to be hurled into "the abyss," Jesus allows them to possess a nearby herd of swine. The swine rush over a cliff and into the sea, where they drown—an odd outcome, since pigs swim well, but that's how the story goes. That last image, the pigs stampeding toward a suicidal dive, seems silly if you picture stubby-legged, pink-skinned, heavy-bellied domestic pigs. But picture instead a fleet, bristling multitude, their jowls flecked with foam, their tusks set at a slight gape for ready use.

The local citizens' committee, hearing of these goings-on from the formerly possessed himself, shows an appalling lack of gratitude by asking Jesus to leave town.

The madman of the tombs and his companions, the swine. The story influenced Western thought for centuries. In medieval Germany pigs were subjected to exorcism before they were slaughtered. Even today people claim to be possessed by demons or multiple personalities, and the script they follow—the priest demanding names, the torment of the raveling truth, the frantic bargaining of the evictees—is the story of the swineherd.

To understand the pig, we should now take a long detour into the lives of insects and salamanders.

One warm spring day during World War II they brought the snowplow out in a little Ohio town. It pushed great drifts off the streets and onto the sidewalks. The drifts were thickest beneath the streetlights, some of them three or four feet deep. They weren't made of snow.

They were made of mayflies. Each was brown, slightly furry, with transparent wings that jutted above its back at forty-five degrees when they were at rest—though in death they broke off and scattered everywhere; you found them sticking to windshields and freshly washed dishes. If you had seen a mayfly in the day or two before the swarm, you might have taken it for a mosquito, if you paid any attention to it at all. They are plain-looking creatures made noticeable only by their quantity, though water pollution now thins their numbers and prevents the great swarms that used to occur in the Midwest.

A mayfly spends its youth as a wingless carnivore before it crawls up a stalk of grass from the bottom of a creek. It hangs there drying until its back splits open and a winged insect painfully extracts itself, leaving its old aquatic self a withering shell. The winged thing crawls up the stalk a little farther and clings there, still as the dead, through the cool night and the following morning. Then it, too, splits a seam along the back, and a fully developed mayfly emerges.

Through a seemingly miraculous bit of timing we don't understand, the mayfly finds that all its fellows have emerged at the same time. The males swarm that evening, forming clouds that eclipse the moon. The females come to watch the swarm and leave with a mate. They couple on the wing. The male is dead before he hits the ground. The female finds water to drop her eggs into. She's dead by the morning, her carcass feeding a fish or cluttering a road.

The mayfly's apparently short life is not so short. The adult only lasts a day or so, but the naiad form, the gilled predator that lives at the bottom of a creek or pond, lives for months or even years. In other words, the mayfly's lifespan is about the same as other insects, but an unusually long proportion of it is spent in childhood.

The same phenomenon of extended youth occurs in various corners of the animal kingdom. Some species of June beetle live for three years as burrowing white grubs before emerging as adults late one spring to fly around eating and mating for a week or two. Some butterflies spend all but a day of their lives as egg or caterpillar.

There's a salamander in Mexico that never sheds its gills and fins for lungs and legs. While other salamanders eventually prove amphibious, this one stays merely aquatic. It even reproduces in its tadpolelike state. So the Peter Pan syndrome in animals doesn't have to include a short adult life following an extended childhood; it can involve a permanent childhood, with the adult sexual powers awakening in that youthful state where they wouldn't normally occur. The principle is called *neoteny*. This broad term includes such diverse phenomena as extended youth and the retention of a single juvenile trait into adulthood, and it can apply to both anatomy and behavior.

The reasons for neoteny depend on the particular case. In one of the butterfly species I mentioned, neoteny seems to occur because the predators in that ecosystem, mostly birds, eat a lot of flying insects, but not so many crawling ones. By reducing the time it spends as a vulnerable flyer, the butterfly improves its odds of surviving. The human form is basically that of an infant ape. Our big toes don't rotate into the opposed position to give us the grasping feet of adult apes; they stay parallel to the other toes, the position seen in a gorilla fetus. This trait makes us better at walking over open ground. We also have long childhoods; we reach puberty later than apes do. This apparently allows our brains to grow larger, and a larger brain results in extra intelligence.

Neoteny affects many mammals. For example, it's normal in felines to be trusting, friendly, and curious as infants. Trust makes the kitten or cub capable of accepting parental care; a friendly demeanor helps the parent want to care for the cub; and play allows the physically underdeveloped animal to develop hunting skills by practice. Most wild members of the cat family grow out of these traits, because they soon outgrow the need for parental care and education and are ready to live a solitary life in which having other big predators around is no advantage except at mating time. That's why so many people who buy a young tiger or mountain lion as a pet get attacked when the cat reaches breeding age. No longer in need of a surrogate parent, the cat begins to see the human as an enemy.

But domestic cats usually retain their childish features into breeding age. They continue to like being touched by their human companions, whom they treat as parents, and they return the affection. They like playing with strings and other toys all their lives. This neoteny is a survival trait, because it allows cats to live symbiotically with humans and therefore to be fed reliably.

Neoteny doesn't have to affect the entire animal. It's possible for an animal to have one juvenile trait in an otherwise standard adulthood. Sometimes neotenous traits appear in unexpected groupings. Researchers who tried to domesticate white foxes to simplify the harvesting of fur were stymied by one such grouping. They found that the juvenile trait of docility, which they could produce after a few generations of selective breeding, was accompanied by the mottled coat of a young canid. Of course, a mottled coat defeated the purpose of using the animals for their fur.

We don't understand exactly how neoteny works genetically. Some biologists think there's a special gene for neoteny which controls how other genes express themselves. Whatever the mechanism, the fetal development of an animal group provides it with a treasury of resources that can be exploited when necessary. Environmental pressures bring out some helpful adaptation derived from some early stage of the animal's development. This happens in a few generations, a much shorter span than that required for most evolutionary adaptations.

Neoteny is a component—maybe the main component—in domestication. Most domestic animals reach adulthood with a mix of what would be adult and juvenile traits in their undomesticated counterparts.

We share some neotenous traits with domestic pigs. A wild boar is hairier than a domestic pig; our closest relatives, the great apes, are more hirsute than we are. Domestic pigs and humans both have relatively short snouts, as wild piglets and baby chimpanzees do. I'm not sure these traits have any particular advantage for either of us. Other neotenous traits, however, definitely affect the pig's status as a domestic animal. For example, the ability to mate and conceive at any time of the year is characteristic of pubescent mammals, rather than adults. The domestic pig is useful to us as food because it reproduces far more often than wild pigs with breeding seasons, and can therefore supply us with more meat.

The pig's neotenous ability to retain fat—literally baby fat—also makes it useful to us as food. Wild pigs are leaner. ("There's hardly any meat on a piney-rooter," a former pig farmer told me, using a local term for feral pigs.) We even enhance the extended youth mechanically, by castrating young males.

When domestic pigs escape, or are turned loose to build a wild population for hunting, something strange happens. After a few generations, the pigs look like wild ones—the long legs and snout, the color, the hairy hide. They have reverted. It's precisely the opposite of the domestication process, in which those wild characters are extinguished (or rather, submerged to smolder) in favor of fat-building body styles. The changes we make in swine undo themselves in the wild.

There are myriad myths from almost every culture that claim modern man is inferior to some earlier race—giants, angels, gods, men of gold. There are fictions, from *Tarzan of the Apes* to *The Jungle Book* to *Lord of the Flies,* that wonder how wild we could get if we were separated from our own kind. There are science fiction stories in which our precarious civilizations collapse, some catastrophe turning us loose from ourselves. These are all, I think, reflections of our deep unease. We know how close we are to the wild. Much of our success comes from neotenous abilities: varying our body fat with the climate, reproducing without season to stay ahead of our predators and competitors, leaving our skulls unsutured at birth so our brains can balloon. We have not exactly evolved from the ape; we are apes whose changes, whose minds, are made of childhood dreams.

———

Witness the dissection of a human body. Try not to notice the surface of the body: the curve of the nose, the color of the hair—in short, the characters that mark this man as an individual. Do not see him that way, because we are about to delve beneath the surface.

The first incisions: a diagonal across each breast, then a long one that begins just below the sternum and ends at the pubic bone. The opening of the ribbed flesh with shears and retractors—a surprisingly loud, untidy process. A tour of the organs: the lobes of liver, the veins covering the heart like ivy, the mass of the pancreas, the single stomach, the haphazard loops of intestine packed in a filmy membrane, the kidneys you have always pictured as distinct little beans but which are really mere protuberances on the back of the body cavity.

Now witness the slaughter of a hog. The iron jolt as twelve-pound sledge meets skull: more a vibration of the ground than a sound. The thin knife finding an artery in the throat, an artery whose analog right now shudders against your own voice box. The blood erupting in crisp jets; the hog hoist by its hind feet, the wait for the last tangled stream of the blood. A winch dips the carcass in a barrel of hot water, preparing the hide for the scraping that removes the hairs. A cut down the middle to open the viscera, the stroke of an axe to break the sternum and let the ribs swing out like saloon doors. The rest is familiar from the cadaver: lobes of liver and the rest, even to the single stomach in its caul of fat—not several stomachs, as horses and cattle and even the pig's cousins, the peccaries, have. The intestines in their filmy membrane look, as human intestines do, like long sausage casings—of course, these will be. As the guts come tumbling out together (close the intestine with a string and cut the anus out to keep the filth clear of the meat), look for the kidneys, stuck on the back wall of the body cavity.

The second-century physician Galen looked into the body of the pig to know the human heart—the law forbade him to work on cadavers. Galen's findings formed the unquestioned laws of anatomy for over a thousand years. The age of empirical science transformed the human way of thinking, revealing the lies in many things we thought we knew, but it didn't change our model of human anatomy: we still know it to be close kin to the pig's. Anatomists call the pig "horizontal human" because of what we share inside, just as the human cannibals of New Guinea called their victims "long pig."

The resemblance crops up in a hundred uses. People training to be forensic entomologists are presented with murdered pigs wrapped in sheets or buried in shallow graves, and asked to find the time of death. They deduce as if the corpse were human, because the same insects recycle us both. Experimental pigs are murdered, dismembered with axes and saws, thrown into condemned houses, and the houses burned down. Doctors study the bones to know how the evidence of murder alters in fire, because our heated bones go through the same shifts of color as the pig's. Researchers try to find ways to pack more meat onto the frame of the pig. They've been splicing genes, adulterating the pig with the only domestic mammal that's better at packing on pounds: the human. Doctors need proteins from human blood to treat hemophilia and heart attack, and gene splicing can make a pig's milk rich in these proteins. Maybe you dissected a fetal pig in college biology and saw the human anatomy in miniature.

It is the kinship of pig and human that makes their meat dangerous to us. The parasitic worms we get from them are human parasites, equally at home in a horizontal or vertical gut. Our domesticated proximity, our insistence on feeding them our refuse and making them live penned in their own dung, gave them these gutworms. They are like us inside, and also out. The array of pig colors includes smooth black, chocolate, rich tan, the sallow of ripe pears, pearl limned with pink—the colors of human skin. A human burned beyond his skin's capacity to heal can be patched with living grafts of porcine hide, which does not sweat but will redden in the sun.

———

We humans eat 88 million pigs in a year, according to one estimate. Such consumption is a corollary of our own domestication, our insistence on settling and having buildings and books. Having outstripped the world's ability to produce animals for our hunting, we must produce our own prey. To manage that, we breed pigs in great numbers and pen them on traditional farms or in mass confinements—"pig factories," as some call them. I wanted to see how a confined breeding operation works.

The idea at the breeding farm was to keep from bringing in disease. We had to shower in, using shampoo and soap supplied by the company. We left our own clothes in the locker room and put on company clothes. A stock of undershorts, coveralls, and boots in many sizes waited on shelves outside the shower. They didn't have my size in coveralls, though, and I had to walk around the place in a Quasimodo stoop.

The pigs were in stalls made of metal fencing, many of the stalls too small for the animal to turn around.

The cement floors beneath them sloped toward drains with metal covers. My guide demonstrated his bravery by putting his foot into the stall of an irritable boar. The high rubber boot that encased the foot was patched with duct tape. The boar snipped; the man deftly yanked his foot back, causing the fence to ring with the impact of the boar's teeth.

I watched the man check the young female pigs, or gilts, to see if they were in heat. Choosing a gilt, the man pointed to a region of her anatomy I had always considered private. "We call that the vulva," he said. Apparently he thought the term had been invented in the pork trade. "You're supposed to use gloves for this, but it's kind of a pain in the ass to find any when you're on the job." He shoved his thumb up to the first knuckle in the gilt's vulva. She shifted slightly, but made no protest.

"Wet," he said, holding the thumb in front of my face so I could observe the mucus. "That means she's in heat. The question is, will she stand to be mounted?" I tried to imagine how he would test her on this score. He opened her stall and prodded her into a larger pen. He came up behind her and pressed the heels of his hands down on the middle of her back. She froze.

"That means she's ready for romance," he said. "Otherwise, she'd have run off when I pressed on her."

He led the gilt past the stalls of several boars and observed which boars got erections. "Let's start with

Ready Freddie," he said to an assistant. He explained that he preferred to use experienced boars who could perform consistently. "Some of the young ones mount from the wrong angle or what have you."

The assistant released Ready Freddie into the pen. Ready Freddie's erect penis appeared to be about eighteen inches long. It was about as thick as a man's finger, and the last third of it spiraled like a corkscrew. It bobbed dangerously close to the ground as he ran toward the gilt. She ran from him for a minute or two before standing firm. He mounted, his cloven hooves daintily folding back on either side of her spine.

The assistant stepped forward to lend a hand. He grabbed the boar's penis and shoved its tip into the gilt's vulva.

"That's one of the really big innovations of the confinement breeding industry," my guide said. "We help them plug in every time. If you were breeding on your family farm, you might just put them in together and hope for results. But the boar only mounts successfully a fraction of the time, and when he manages it, he takes a lot longer than what you've seen here. Our little manual assist raises the conception rate. Of course, some small pig farmers help with the plugging in, but we help with the whole fertility cycle. We monitor the gilt's temperature constantly, so we can get her bred as soon as she hits estrus. It's scientific, not like the guesswork of normal farming." Then, to the assistant, he hollered, "Let's do another for him, Toby."

Toby went to check the other gilts whose temperatures showed them to be near estrus. I saw him walking in a crouch down the row of their stalls, thumb extended. "I have to keep an eye on Toby," my guide said. "He's one of these guys that will rape a gilt."

"Rape a gilt?" I said.

"Force the boar on her, instead of letting her take her own time to stand. That way he can mark down that he's bred her, but it's sloppy work. The stress of it keeps her from conceiving."

———

We've bred pigs for different cuts of meat. We can make them heavy in the rear for plenty of ham. We make them long for plenty of bacon. I knew of a boar eleven feet long and one thousand pounds, so big he broke the backs of two gilts he tried to mount. At a carnival, among tents housing freak shows and fossils, I came across a sign that read WORLD'S LARGEST HOG— 2200 POUNDS. I paid my dollar to get in and I saw him, a Volkswagen Bug of a brute, his back high as my chin, his basketball testicles jutting at the rear. Such a creature could never live wild: he was an artifact.

———

In 1268 a German court tried a pig for killing and eating a human child. The pig, found guilty of murder, was executed.

That was the first court trial of an animal in medieval Europe; others followed. Like the later crusades against supposed witches, the pig trial depended on belief in demonic possession. The courts believed they were trying sentient beings. The New Testament accounts of the swineherd colored the court's reaction to an ordinary act of predation.

That act was probably real. The pig's taste for human flesh is well documented. Historical documents refer to unprovoked attacks by wild pigs. During the reign of Charles I in England, when the Crown shipped in wild pigs for hunting to replenish a forest that had been hunted out, the pigs "became terrible to the travellers."

Every citizen of a rural community knows a "true" tale of a man eaten in a hog pen. Generally, the incident happened "not too far from here"; the man was old; he died of a heart attack or a stroke while feeding the pigs; and "nothing was left of him but the shoes." The repetitive formula of this story and similar ones marks them as legends—you have to be particularly suspicious of the heart attack diagnosed from the shoes. Nevertheless, a rare man-eating story proves well documented. Small children are the usual victims of domestic swine, but the swine take adults as well. Here is a true tale that sounds like a particularly silly version of the legends:

In 1938, in the little town of Harper, Kansas, a man raised hogs for his living. People would see him driving to the grocery stores in his old Model T Ford, which was equipped with a bed like that of a pickup truck. The stores gave him their old, unsalable produce, which he would load into the bed of the Model T and cart off for his hogs. Most people knew him only as Hog Slop Charlie.

A neighbor found a little of him one day. His remains lay scattered in the hog pen. The story most people settled on was that Hog Slop Charlie had died of a stroke in the pen, and his hogs had scavenged him. Of course, no one was there to know; they may have killed him. I have not been able to ascertain what happened to his hogs after he died, or whether anyone ate them.

A friend who fought in Vietnam didn't like to talk much about his experiences there, but he told me that some of his worst memories involved pigs. He mentioned this fact as we stood looking into a pen in his barn. Inside the pen were Vietnamese potbellied pigs, a sow and her litter. The piglets were old enough to root around in the hay. One of their brothers lay dead, half-buried in hay at the center of the pen. His swollen slit of an eye swarmed with blue-black flies. His body ended behind the forelegs. His mother had eaten half of him. He was the runt.

"Pigs eat anything," my friend said.

CANID

The trailer houses were packed close. Some of them hadn't been leveled; most lacked numbers. The residents got their mail at boxes nailed to a hastily cobbled lumber frame down by the paved road. The weed-ridden strips of yard bore accumulations of junk. Everything about the place seemed to say, "We're just doing this until we can get a permanent place."

A cratered dirt drive ran before the trailers. By day, children played there. At night, the dogs took over. They brawled and barked, keeping the human residents awake. They raided the garbage. A few children had been bitten. The ranchers who owned nearby pastures said the dogs were running their cattle lean, and it was only a matter of time until they brought one down.

A minor political squabble developed around the

dogs, the ranchers blaming the owners of the trailer park. That battle was remote from my concern as I sat in one of the trailers one evening, playing cards with some friends who lived there.

A dogfight erupted just under the kitchen window. Dennis picked up his bow and his only remaining arrow and raced to the door.

I caught up to him standing on the front porch staring at the dogs. Their fight had subsided. There were ten or fifteen milling about. A big black mutt lay in the yard gnawing the children's toy lawn mower with the side of his jaw. Dennis shouted a threat. The mutt growled back.

Dennis nocked, drew, and released. The dog ran off screaming. I saw the shaft protruding on both sides of his abdomen.

———

"Buffalo hunting," a friend said when I told him about the incident. He knew two men who hunted trailer park strays as a hobby. They, too, used bow and arrow, since firing a gun in town was illegal. They used the code "buffalo hunting" to conceal their activities from others who wouldn't understand.

"They have a point," my friend said. "When you say 'dog,' most people think of a pet. Those packs of strays aren't pets. They'd take you by the throat if you got too close."

My grandmother had told me the same thing years before as we looked out the window of her trailer at two male dogs battling loudly around the hulk of a rusted-out Mustang. It must have been a fight for leadership. Their pack skirled around them, barking and kicking up dust devils. The combatants, a thin, pied mongrel and a scarred, blunt beast that might have been half boxer, yelped and tried to shield their genitals with their tails, each pushing in for a castrating bite.

———

A man who was training for Special Forces would tell me what he learned. Once it was about sneaking into a guarded building. The human obstacles, he said, were to be handled as one would expect: seized from behind, the throat slit. The canine guards could not be taken by surprise.

"Your big dogs go for the throat. I'm talking Doberman, German shepherd, most of the ones used as attack dogs. You put your arm up to protect your throat. You let him bite your arm, but you fall back with his momentum. As you fall, you put your other forearm just behind his head. As your back hits the ground, you're bringing your knees and feet up to push him up over your head. Basically you're giving him a monkey flip, and you're holding your arms rigid. His mouth is hooked onto one arm, the other's behind his neck, and as he flips his momentum snaps his spine. One dead dog. Not hard to do, but you have to sacrifice your arm. You're okay if you're wearing a thick jacket. If not, your arm gets pretty torn up. You could bleed to death."

———

Killing a pet dog is an act in the realm of the holy. It is so uncomfortable that most people assign it to professionals who will do it out of sight and never make them see the carcass. We tell our children about it in guilty euphemisms. We say the dead dog has been "put to sleep," as though he might wake and release us from the burden of that irrevocable act.

At the other end of this spectrum of dog-love, some believe a man owes his dying pet a death by its master's hand. I know a man who was obliged to kill a dog he loved because it had hurt a child. He took the dog into the country—he wouldn't hear of anyone else doing it. He took a gun and a shovel along, and when he returned he spent a week in a dark and silent mood. His wife says he kept the casing of the bullet.

Farmers and ranchers have codes for dealing with other people's dogs. One man whose bitch had been courted by the neighboring males took offense when a certain male turned up wounded and he was mentioned as the shooter. He might very well have shot an intruding dog, he said; but he wouldn't have unless he were certain of a killing shot. His point was this: A man's property rights outweigh the life of a dog, but no one has the right to make a dog suffer.

A neighbor's border collie once took some of my family's chickens. I would go outside to find a white leghorn lying in the drive, the feathers of its neck frosted with fine drying droplets of blood, and the hen, moments from death, would raise its head to look at me. "I don't have any chickens I can learn him off of," the dog's owner said. "I won't fuss if you have to kill him."

The city dweller's desire to distance himself from the killing of his own dog and the farmer's ethical code for it: both are signs of a deep reverence.

When we hate the dog, we do so not only because it's dangerous but also because it has stepped outside the role we want it in, that of loyal friend to humans. The dog's loyalty is a commonplace, a homily, a fact too obvious for notice. It is also the signpost of an odd quirk of evolution.

———

Gray cinder block walls; oil stains on the cement floor, some of them peppered with absorbent gravel; toolboxes; a greasy workbench mounted with a heavy vise. My father's shop. From a rafter hung a cord that ended in a droplight, which was in my father's hand. My uncle was there too, and they worked together in the pool of harsh light surrounded by darkness, the black hair of their thick forearms gleaming with sweat. Their tools were a length of elm branch and a pair of pliers.

The men occasionally gave each other suggestions. The only other sounds were the snarling and whimpering of the dogs. My father would seize a dog and hold it down, levering its mouth open with the branch. My uncle would move in with the light and the pliers. Jutting from the black-and-pink flesh of the dogs' mouths were the quills of a porcupine—long, flexible needles terminating in splintered fishhooks.

The miniature bloodhound was first. He struggled and growled against my father's grip while my uncle sank the pliers into his mouth. It seemed a tableau of dentistry in hell. The other two dogs moved to the edge of the circle of light and stood with their tails alternately between their legs and wagging in appeasement. I thought, why do they stand there waiting for it to happen to them? I knew it was supposed to be for their own good, but how could they know?

In the morning my mother showed my sister and me a little plastic pumpkin—it was nearly Halloween. In the pumpkin were a dozen brown-and-white quills. When we went outside the dogs were happy to see us, as though everything were the same.

———

One thing the dogs did for us was to reveal the other life around us: the nocturnal animals we rarely saw, the burrowing animals only they could hear and smell, the distant things making sounds imperceptible to us. We rarely saw raccoons, until the dogs showed them to us. Raccoons weren't especially common in our part of the state, where trees were scarce; we only saw them occasionally on a long drive home at night—the sudden fire of fractured amber that must be the eyes, a thick grayish body bounding across the dirt road, fast enough to blur, though the headlights somehow made it seem suspended in the dusty air.

The dogs would scent them at evening, and a chase would begin—the raccoon somewhere ahead, beyond my sight; the old dwarf bloodhound leading the pursuit, his crumpled lip and throaty growl announcing his intention to kill; the lean, lupine border collie at his hip, faster than the leader but lagging to show respect, barking to keep the slower animals apprised of the quarry's location; the indefatigable short-legged mongrel with the skunk tail a few steps behind, also barking; my older sister a few yards behind, covering ground with her long-legged strides, sometimes followed by a few cousins or friends; and me, the slowest of the pack. We humans, even at our age, thought the dogs were serving us. I wonder how we held this conclusion as we labored to keep up with them on a hunt the bloodhound had devised.

Such hunts happened dozens of times in my childhood, and the archetypal example that has crystallized from the mix of details in my memory involves irrigation pipes. These aluminum tubes, stacked like cordwood, each big enough to hold a human head, always seemed to be lying around somewhere waiting to be installed; I don't know how the crops ever got irrigated. The quarry—it might be a jackrabbit or a cottontail, a raccoon or a possum or a porcupine—dashed into a pipe. The dogs rushed along the open ends of the pipes—not looking, but smelling—until they had found the one that sheltered the animal. If they couldn't find it quickly, they would pause in their sniffing. The mutt and the bloodhound had floppy ears; I could see the cartilage of those ears moving slightly. The border collie's ears would abandon their usual submissive backslant and stand up, rotating to triangulate.

That did the trick: all three dogs suddenly knew exactly where the prey was. They split up to cover both ends of the pipe. Then they barked into it, their manic cries echoing from the sheet metal buildings nearby. They pushed their snouts into the pipe (the bloodhound always first) and their cries were trapped in the pipe; we could touch the pipe and feel it vibrating with the sound. I imagined the raccoon curled in the middle of the pipe, as distant as possible from both ends, his gut shaking in time to the threats against his life.

The raccoons were always smart enough to stay in the pipe and wait, even through hours of barking. We poked at them with sticks and hurled rocks into the echoing pipe. Once, early in the morning, I found a sparkle of blood on the lip of an irrigation pipe among the beads of dew. The raccoon we had harassed for ninety futile minutes the night before had finally walked away during the night, after we had all, canine and human, lost interest. He had taken some sort of wound during the harassment, but had sat still for his only chance of surviving.

The possums also knew they'd better not stop to fight the dogs. They would curl up and cling tight to something inside the pipe. We couldn't move them an inch even by tilting the pipe. Once it was an albino possum. As he reached the pipe ahead of the bloodhound, he looked back at us with his mad crimson eyes, baring his milky teeth and hissing. He looked like a demon with leprosy. Possums will sometimes back up such threats with action, but this one must not have liked the odds. He disappeared into the pipe.

Rabbits could sometimes be spooked out of the irrigation pipes; they were faster than possums and raccoons, but, judging from the number the dogs caught, I think their tendency to abandon cover was really a sign of lower intelligence. The rabbits would run in zigzags, a strategy for evading single predators that rarely saved them from the three dogs giving chase as a team. The old bloodhound would put on a burst of speed, prompting the rabbit to spring sideways. The other dogs would be running behind and to the sides; the rabbit would often jump directly into the jaws of one or the other. This dog would seize the rabbit by the neck. Then the bloodhound would rush in snarling, take a mouthful of rabbit, and pull away from his colleague. As a child, I thought the older dog was trying to take the rabbit away from the younger. Years later, when I saw other groups of dogs hunting in the same way, I realized that this pattern is actually a killing tactic.

The second dog to score a bite, usually on the rump, tail, or hind leg, pulls away from the first to stretch and bend the prey's spine. Both dogs shake their heads, which makes them appear to be wrestling each other for the prey; maybe they really are. But they're also cooperating to break the spine. If the first biter has a grip on the throat, the dogs' pulling away from each other strangles the prey.

Another advantage of the dogs' tug-of-war is defensive. When stretched, a prey animal has less chance of clawing or biting the dogs. As the dogs continue to pull, their tug-of-war makes their teeth tear the flesh. Other pack members try to get in a good bite. Sometimes you can see the struggling dogs working their jaws around on the prey, finding a grip that will drive deep into soft flesh. The prey bleeds until it falls into shock. It is defenseless and ready for death.

In India, the dog of the streets belongs to no particular breed. It is a tan, short-haired, generic-looking creature, and is generally held in contempt. Where the human population thins a bit, the dogs (they are called dholes) run in wild packs. They snip the sides of fleeing deer until the guts are exposed, then latch onto a bite of intestine and change direction. The bowels come raveling out; like the human body, the deer's contains a startling yardage of gut in a small space.

It is a painful way to die, as Elizabethan annals of crime and punishment testify. Traitors were hanged for a few minutes of kicking strangulation, then cut down, still alive, to watch themselves castrated. Finally, the torturer (it was a profession of sorts) sliced into the traitor's belly to reveal the intestines, which he grasped with tongs and drew out. This invariably proved fatal, though usually not quickly. The purpose of this regimen was maximum pain.

It is often the way of death for victims of the dhole— not just the drawing, but also the castration, which the dhole's jaws may accomplish during the chase or after. If events fall mercifully, the jaws may strangle the prey to death early on. Or the animal may be brought down by the drawing, to watch as the pack eats him alive.

The dog of the Middle East also is a pariah, a wild scavenger usually not attached to particular humans. It is regarded with revulsion at least as ancient as the Old Testament, which declares the dog an unclean animal, not fit for eating. (If you're surprised the question even came up, consider the Aztecs, who fattened dogs as livestock before eating them.) It serves humans in a detested, but useful, role, as a roving garbage disposal. Meat unfit for human consumption is tossed to the dogs, as are table scraps.

The Bible persistently uses the dog as a metaphor for contempt. The starving beggar Lazarus is so low even the dogs lick him. Goliath thinks David shows him the disrespect owed a dog by coming armed only with a sling and stones. Even when doing some useful job, like guarding a flock, the dog is held in contempt, its name consistently linked with dung or vomit. It is also feared as a pack-hunter not averse to taking human prey.

The dog's poor reputation in the Middle East stems from the same source as the pig's: both animals scavenged human corpses. The Hebrew God's punishment for rulers who defy Him is that after they and their families are killed, their corpses lie unburied. The Old Testament describes dogs and birds eating such corpses. The Bible repeatedly refers to dogs licking up human blood.

The biblical Jezebel is thrown from a high window to shatter on the flagstones below, and when the men go to dispose of her corpse, they find the stray dogs have left only her skull, her feet, the palms of her hands.

———

When I was a child, the predatory tactics of our dogs made me think about the differences between wild animals and domestic ones. So did the long, bloody leg bones of cattle we sometimes found in our yard, the dogs snapping at each other briefly to establish the right of gnawing out the marrow.

But I really got interested in the wild and the tame at bedtime, when the howling began.

It would start somewhere out on the plains, a sound like the otherworldly cry of the owl compounded with unspeakable loneliness. It would double, as if the same coyote suddenly stood in two places; choruses would join in; voices would compete to scramble over each other, then suddenly fall into harmony. Then, close to the house, the dogs would howl, each dog's voice recognizable and distinct.

I'd seen a few coyotes by day. The dogs would suddenly stop whatever they were doing and stand gazing toward something unseen in the prairie grass. The bloodhound would bark—low, inquisitive, with a touch of menace. All three dogs would burst into a run. The coyote would appear far away, bounding in and out of the grass as he ran, the dogs pursuing. The coyotes were fast. Their appearance was only a second ahead of their disappearance. Sometimes a coyote came for the chickens in the night. My father would rush out with a rifle, warned by the dogs.

So coyote and dog were enemies. Yet I had seen a dog in town that my grandmother pointed out as part coyote. My uncle had shot a raiding coyote once, and when we saw the carcass up close it had long reddish hair like a collie: further evidence of crossbreeding. And when coyote and dog howled together in the night, it didn't sound like an exchange of threats. It sounded like a shared song.

———

The coyote can live where people can't, and even where wolves can't. He inhabits the desert, but also survives in snow. He dwells in the banks of rivers unseen by human eyes, but also walks into cities in daylight to scavenge human garbage. The coyote flees from our gaze, but never leaves our lands. He comes back when we sleep to take a lamb or a calf. He raises his cubs within our hearing. He knows us, and fears us only within reason. He knows our guns kill only along the lines of sight.

In Spanish his name has come to mean "crafty." He knows how to kill rattlesnakes without getting hurt. He knows how to catch rabbits by driving them until they hit their territorial boundaries—a male rabbit will not intrude on the territory of another male, even to save himself from being eaten. The coyote tracks wolves to scavenge their kills. He doesn't eat the kills of cougars unless he's desperate, because cougars are more likely to track him down and take revenge.

Coming from a country graveyard one cool Memorial Day, I saw a coyote on a mudflat near a streaming ditch. The face and legs seemed delicate, as if they might prove smooth and ceramic to the touch. The body looked too thick for those slim extremities, encased in too large a coat, and the tail hung back and down, thick as a haunch. The coyote did not break into undulant bounds, or jet out, but walked away, the slender legs working in rapid, perfect coordination, like an insect's, the furred body gliding forward. A second coyote I hadn't noticed followed the first one into the grass; a third turned to look at the car before he joined them.

———

The coyote's equivalent in the Old World is the jackal—some jackals are indistinguishable from coyotes by sight, and they may really be the same animal. The biblical abhorrence of the dog extends to the jackal. In the Old Testament, he wanders the streets of ruined cities, a metaphor of desolation strangely echoed in the Egyptian god of the underworld, a man's body with the head of a jackal. Job evokes his desolation by claiming brotherhood with the jackal. (Some translations confusingly call the jackal "dragon," a term also used for some great reptile.) In Revelations, the jackal is the mother of the Antichrist, the root of apocalyptic evil.

———

I came to Philip and Alberta Hart's house to see captive wolves. As I reached the front door, I glimpsed them in the yard: three white wolf-dog hybrids and a pure white wolf, blizzarding around each other behind the fence. Their tongues lolled and their tails wagged before they whirled out of sight.

Philip Hart took me into the yard to see them up close. The pure wolf, a waist-high male, ran at me. He was a lithe seventy-five pounds, not especially large for a wolf, but bigger than most dogs. I offered him a smell of my closed hand, as I would a strange dog. His perfect teeth were white, not even slightly yellow, and his coat was an unbroken white too. He took my hand in his mouth and held it there. The point of one long canine tooth rested on the knuckle of my forefinger, and further back another tooth clicked against my wedding band. I told him how handsome he was. His mouth had no trouble accommodating my fist. I took a good look at the carnassial teeth on the side of his jaw, the specialized ones with which a wolf shears meat from a carcass. I didn't think he would have much trouble removing my hand.

"He's being friendly," Hart said. "If he wanted to hurt you, he would have come at you with his mouth closed." The wolf released my hand and put his fore-paws on my chest, standing on his hind legs to look me in the eye. I thought I should have been terrified, but I wasn't. I could read the wolf's body language. It was the same language dogs use, a language people read with-out thinking about it. Posture, tail, ears, and hackles had told me I was never in danger.

Eventually I got around to asking my host about the danger of wolf-dog hybrids. He raised and sold hybrid pups. Hybrids had been in the news because a few had killed human children. States were passing laws to limit the percentage of wolf blood allowed in a pet.

"That's the wrong idea," he said. "People think the hybrid goes crazy and kills because the wild wolf genes eventually express themselves. But wolves don't kill people; dogs do." He was right. In the United States, domestic dogs kill more people than rattlesnakes, ven-omous spiders, stinging insects, bears, or sharks. Dogs, in fact, are second only to humans as killers of humans. Pure wolves, wild or captive, kill people so rarely it's hard to find authentic cases. Of all the dangerous ani-mals that are supposed to fear humans innately, the wolf is the one whose fear has been documented. Scien-tists tracking the movements of wolves find they some-times abandon great tracts of land so they'll never have to cross a human scent trail.

"The hybrid has the wolf's instinct for establishing its dominance, plus the dog's lack of fear for man," Hart continued. "The more wolf you put into a hybrid, the safer he is. The closer he is to fifty percent, the more likely he is to kill somebody. I don't even do many fifty-fifty crosses anymore. My crosses are wolf father, hybrid mother, so you get a high percent wolf."

———

More than a century ago, Native American and European American were spilling blood for the Great Plains, a country whose plant eaters were culled by wolf and cougar, black bear and coyote, eagle and owl and rattlesnake. The land was pocked with the vast cities of the prairie dog, cities whose population could run into the billions. But the animal that dominated any traveler's sense of the Plains was the bison. It moved in uncountable herds whose wallowing and walking and pawing shaped geography, whose thirsts controlled the surface water as much as storm and stone did, whose carcasses sustained predator and scavenger.

A bull bison is massive, providing enough meat to feed a tribe for days. It's powerful enough to pulverize an intruding human. Back then, people would occasionally see, among the bison herd, one male far larger than all the others. This monster, when brought down for his hide or his meat, would prove to be a eunuch. Castrating a male bovine makes him grow bigger. That's why beef sold in stores comes from steers, which are castrated males.

The giant bison were survivors of wolf attacks. A single nip during a chase was enough to sheer the testicles off.

When a wolf pack tackles a large herbivore, there is a pause before the chase, a moment when potential killer and potential prey eye each other, as if making a pact to play their roles. The herbivore that buckles and runs has signed a contract.

Wolves can run for miles. They take turns resting while pack mates harry the prey. This tactic multiplies the wolf's individual stamina, so that few herbivores escape a wolf pack by tiring it out. The pack can take down a moose or a bison when it's exhausted and some of the wolves are fresh.

A favorite killing move for wolves is to tear at the windpipe. Another is to spill a lot of blood. As the herbivore runs from the pack, he leaves his vulnerable sides and rear open to quick bites. The bite of a wolf, even on the run, does heavy damage, and an accumulation of bite injuries drops the prey: shock, caused by loss of blood.

———

The house had been expanded down the mountain. I descended sets of steps between rooms. There seemed to be a lot of rooms. Finally I reached a sliding glass door, and out the door I saw two wolves standing in the dark.

I had come to visit Bud and Nancy Saunders, who

kept the wolves. Moments before I arrived, the couple said, a small herd of deer had crossed the pasture, and the wolves had howled. Nancy handed me a tape recorder cued up to an old recording of the wolves' own howls. Play it to them, Bud said; maybe they'll talk back. I pressed PLAY.

The sound from the recorder was thin, mechanical. The echo that broke from the wolves' throats to envelop the smaller sound was fluid, beautiful. It woke in me something close to fear, but quieter. It was like the coyote song I remembered, though fuller and deeper. It seemed to have a current of meaning flowing just beyond my grasp.

When they let the wolves in, I sat on the floor before the fire, which brought auburn highlights out of their gray-brown overcoats. I scratched each wolf on the chest. Their coats were deep. I would lose my hands in them and bring them out slick with the oil and smell of wolf. When I got involved listening to Bud and Nancy, I stopped scratching one of the wolves. She wrapped her huge paw around my wrist and pulled. I resumed scratching. Their paws are more prehensile than a dog's, Nancy confirmed. They clamp their front legs around the necks of big herbivores and score the flesh like cats.

I looked at a wolf and saw something I'd never seen in a dog's eye: my reflection. "That's because she's looking you in the eye," Nancy said. "Dogs don't. They're submissive to humans. Wolves aren't, not permanently."

"These aren't pets," Bud added. "Dogs are pets. They think you're the master. Wolves think you're a pack mate. Leadership in a pack is always open to challenge."

I asked them if they'd ever been challenged. Nancy told about a time she'd fallen in the presence of the wolves. One of them came for her. Nancy came up slugging. The skirmish was over in a second, with no one really hurt. Nancy said if she had hesitated to hit the wolf, things could have been very different. Some wolf owners make a policy of thrashing the animal thoroughly the first day he comes to live with them, just so the hierarchy is clear.

There's a pattern in the attacks of captive wolf-dog hybrids: they go after children who fear them, or adults who show weakness—a limp, a stumble, anything to tell them their time to rise in the pack hierarchy is at hand.

———

We have always forced the other inhabitants of our planet into roles of our own scripting. The scripted roles come in obvious packaging, like the biblical portrayal of the snake or the moral-laden animals of Aesop's fables. They also come in subtler forms: every encounter with an animal holds shades of meaning cast not by the immediate circumstances, but by books, folk-lore, even the tired metaphors of our language. In fact, we interpret most of our interactions with animals sym-bolically, inflecting them with anthropocentric emotion. I know men who pursue the killing of snakes as if each one embodied satanic evil. When I trap a mouse in my house, I feel as if I've triumphed over a burglar. Surely my pleasure in watching a hawk has something to do with the freedom I half-consciously allow it to symbol-ize.

The wolf is embedded in human history so deeply his truth may never come clear. In Western perception he has been demon, devourer of human flesh, and raider of stock. He has also been an alter ego: the werewolf as emblem of human pleasure in sin, the berserker as em-blem of human prowess in war. Americans conflate the wolf with wilderness and Native peoples to symbolize noble savagery, a myth that has antecedents in the wolf-suckled builders of Rome, in the stories of feral children nurtured in lupine dens. For some people, saving him from extinction is a step toward redemption for our ecological and genocidal sins. I often notice such people disparaging the coyote in favor of the wolf, as if the smaller canid's continued success, his talent for exploit-ing human proximity, were a sellout—to whom or what is never clear.

The wolf's power to make us imagine is a function of taboo and danger. Our ancestors saw him among the corpses after battles, eating human flesh. He learned from that our darkest secret: that we are good to eat. But that was only a rediscovery of a fact he knew before we learned to write things down.

———

Species distinctions among wolves, coyotes, and jackals are often based on size—the coyote is smaller than the red wolf, which is smaller than the timber wolf. The standard species divider for animals in general is inter-breeding: if two animals can copulate and thereby produce fertile offspring, they are of the same species. Horses and asses are different species because, although they can produce offspring together, the offspring, mules, are usually sterile. With the canids, this test yields confusing results. Wolves, dogs, jackals, and coyotes can interbreed into many different hybrids. Some breeds of dog are even said to cross with foxes.

The canids challenge our concept of species. Intuitively we perceive them as different from each other, but their frequent interbreedings mark them as identical species. Meanwhile, the dog, which we accept as a single species, shows more range of body style and behavior than can be found between, for example, wolves and coyotes. There are tiny dogs bred to sit in idle laps; there are huge ones bred for killing lions, boars, tigers, and wolves. There are swift ones bred for hunting coyote: they stretch and savage the quarry as other dogs would a rabbit, but they don't eat him. He's too much like a dog, and even his butchered and disguised flesh seems to invoke a canine taboo.

We used to breed great mastiffs to carry harnessed lances and pots of burning tar among enemy armies. Caesar's spirit will "Cry 'Havoc!' and let slip the dogs of war," Shakespeare's Mark Antony says, not altogether figuratively. Dogs are bred for everything from hauling to scenting to herding to burrowing after badgers. Their bodies have taken diverse shapes to mirror their jobs.

The boundaries of canine species constitute a mystery in which we are intimately involved, and its roots reach into prehistory.

————

When predators come out of the trees, they change.

There was, forty million years ago, a primeval carnivore, ancestor to the ones we know today, the animals like wolf, tiger, otter, wolverine, skunk, and all the others that shear the meat from the bodies of other animals with a scissorslike modification of the side teeth. The primeval carnivore lived in jungles, climbing trees with its formidable claws, finding prey with its sharp senses and its stalking skills. Its evolution took two different paths. One branch stayed in the trees, maintaining the solitary life of a climbing hunter that can still be seen today in the leopard and the feral cat.

The other branch came down to the ground, where it eventually diverged further to fill many ecological niches, from the omnivorous lifestyle of the bear to the semiaquatic habits of the otter.

But the divergence from this branch that interests me here is the one that became canid. The wrists that had flexed for climbing in the primeval carnivore stiffened for running, because the canid line specialized in running down prey. That specialty affected more than the structure of the limbs. It led to cooperation in hunting. Stalking had worked in the trees, but on open savanna the chase was everything, and predators in groups could chase more effectively than solitary ones.

Cooperation in hunting led to an elaborate social structure. Wolf society is built around extended families: a dominant couple of breeders, a system of care in which the young grow into supportive subordinates or else leave the pack to join another, an elaborate body language, a vocal language, gestures of appeasement and unity and anger.

The same pattern has recurred several times. The cat line diverged again later to produce the hyenas, which developed into running, gang-fighting predators with an elaborate social system. Later still, another group of cats, this one staying anatomically close to the ancestral line, came down to the savannas. The lion is a social pack-hunter, unlike other cats.

The cat lifestyle never ceased to be workable, though.

The fox rediscovered the ways of his precanid ancestors. His behavior is more like a cat's than a dog's: solitary except to breed, a stalker instead of an endurance runner, a climber of trees, an eater of mice and grubs; his tiny kits are the prey of the owl and the eagle.

The general canid pattern is pervasive enough to crop up in other, unrelated lines. Baboons have not only the social structure but even some of the anatomy of dogs, to which they aren't related—an example of convergent evolution. The obvious case among primates, however, is the human line. We, too, descended from trees, became hunters of fleet ungulates, and consequently developed a complex social system.

The similarities between human and wolf are profound: vocal language, complex society, body language so similar we can read each other, a deep predatory pleasure in killing that causes both species to sometimes slaughter far more than necessary. The affinity runs deep in our patterns of development. Both animals have clumsy young with large heads and feet. Adults respond protectively to children of their species, but the cues that cause the response are general enough to transcend species. That's why humans are capable of adopting puppies. There's evidence (though disputed) that wild wolves sometimes protect human children, presumably because of the same broad instincts.

Wolves and humans both go through phases of intense learning in particular areas. Certain tasks have to be learned during these "imprinting" phases, or they will never be learned at all. For wolves, hunting is an imprintable skill. For humans, language is one. But a canid's hunting instinct can, during imprinting, be modified to fit the society it's in. For example, it can be trained to serve human wants.

These affinities make it possible for human and wolf to mingle in a hybrid society. One more ingredient completes this symbiosis: neoteny, the retention of juvenile traits into later life. For the dog, as for the pig, neoteny is an evolutionary shortcut. It allows the animal to pluck adaptations from anywhere in its individual development, without having to evolve them gradually. It's what separates the dog from the wolf. The dog is a perpetually childish wolf. He fits into human society by retaining his infantile desire for constant affection, making humans love him. He retains also his youthful submission to authority.

More important for man's interest in the dog as a living tool, neoteny opens an enormous range of physical potential, a plasticity inherent in the changes mammals undergo as they mature. Imagine your own body had become stuck at the age of thirteen, when you were having a growth spurt. You would have grown larger than most humans. Neoteny allows dogs, with a few generations of selective breeding, to get smaller or larger. But size is only one example; any trait that changes with stages of development can become "frozen" at some stage. That abnormal development can be passed to offspring. We can therefore breed dogs that are, for example, fiercer or more dependent. Most specialized dog skills have some neotenous component. In short, neoteny gave us different breeds of dog.

When dogs go wild, they tend to be skulkers on the fringes of human settlements, living on our trash. They are more like juvenile delinquents than independent adults. They lose the specific breed qualities we inculcate and become generic, like teenagers growing away from their parents before forming their adult personalities.

––––––––

We humans, as I've said, are domesticated too, though we think we are masters. Hundreds of nights I have trudged out, snow or flood, to feed and water the dogs: I serve them. I have staggered with heavy buckets to the hog feeder: I serve the swine. I have curried the burrs from the hide of a horse, brought bread to migrant ducks, scattered feed for chickens, shoveled dung from the pens of cattle: these are all services I perform for "lower" animals.

The care of animals, along with the tending of crops, is a root of our social structure. It dictates our need for permanent homes, our construction of walls and fences, ultimately our economy and culture. The dog makes this possible, because it was the dog, with his keener nose and ears, that made it feasible for us to protect livestock from nocturnal predators. Our tools, intelligence, and eyesight complement his senses; we share a territorial instinct that gives us a common goal.

Early European explorers of North America sometimes found themselves trailed by lone wolves who scavenged their garbage and woke them to passing bears or strange wolves. This must be how the partnership of human and canid started in prehistory. Probably humans eventually took the pups of wolves and jackals from dens and raised them. This partnership caused the divergence of dog from wild canid by putting certain wolves and jackals in the domestic situation that brought out neotenous characters over the generations. In short, the difference between a wolf and a dog is the human touch. It is not a case of conscious control. As a fungus captures an alga to form the symbiote known as a lichen, we have captured the dog in a symbiosis that has remade us both.

———

A human who decides to cross dog with wolf is also trying to cross an ancient rift between tame and wild, between what lives with us and what moves unseen in the night. We made that rift thousands of years ago when we took in the pups of wolves and jackals that scavenged our middens. We deepened it when we set these adopted canids against their wild brothers, girding them with spiked iron collars and teaching them to protect human habitation and stock from their own kind.

Ancient as that rift is in human terms, it is young in the longer perspective. The earliest evidence of cohabitation between human and dog is no more than fourteen thousand years old: a mere wink before we began to record our own histories, a fraction of our existence as social hunting animals of the savannas, which measures in the millions of years. The relationship progressed differently from place to place—the dog revered in northern Europe, reviled in Palestine. But those differences reside mostly in our perception. Dogs everywhere are tangled in the conduct of human life, as the lowest members of our societies.

Everywhere in the world, at virtually the same time, human and wolf formed a partnership. That is the strangest part of it all: the sudden universality of a bond across species. This bond distinguishes the dog from other canids. It also distinguishes modern humanity from its older branches, because it is an essential element of the change from hunter-gatherer to the settled life.

If we humans died out tomorrow, the dogs and wolves would begin the healing of their long rift. The dog would bring its adaptability and genetic diversity to the wolf; the wolf would bring its independence and social structure. The special breeds would disappear in a few generations, the weak hearts of the pure dying out in competition with the vigor of the mongrels. The jackal would slouch through the streets of our ghost towns.

R E C L U S E

Syndrome 1: The Rotting

A friend of mine, an amateur herpetologist, owned a Haitian anole, a lizard about half a foot long from snout to vent, with another three inches of tail. The anole escaped in the man's house. The man found and recaptured the lizard within a couple of days, much to the delight of his family.

He noticed something strange about his pet. There were two wounds just inside its mouth, and they quickly developed into abscesses. A discolored knot appeared over the anole's left eye. The man guessed from the mouth wounds that the lizard must have eaten something that injured it. He searched his house and found a number of brown recluse spiders, but no insects, no other arthropods of any kind. The brown recluse is smallish, dull in color, plain except for a mark on its back that gives it the common names *fiddleback* and *violin spider*.

The man knew what the brown recluse and its cousins were capable of. (There are, depending on whom you ask, about a dozen recluse species in the United States.) He deduced that the anole had eaten one of the spiders, which had bitten the lizard's mouth on its way down.

Soon the anole's symptoms gave him gruesome confirmation of his diagnosis. The knot over its eye turned into a soupy mass of gray-brown dead tissue—a necrosis, as the scientists call it. The wounds inside its mouth grew and softened into cheesy masses as opportunistic fungi infected the dead tissue there. The anole was dead within two days of its recapture.

Several animals react as the anole did. Rabbits bitten by the recluse invariably develop necrotic lesions, and sometimes die from them. The same is true of guinea pigs. Scientists use rabbits and guinea pigs to research the recluse's venom because of their consistent reactions.

———

The woman noticed a pain on her calf. It felt like a bruise. The area turned red and began to swell. A black point appeared in the center of the red. The pain was bad enough to send her to the emergency room, where the diagnosis was "loxocelism," a condition caused by the bite of a recluse spider. The woman had been working around her house and garage, but she had no idea precisely when or where she'd been bitten.

The doctors couldn't do much for her. The antidote to recluse toxin works only if it's administered within half an hour or so of the bite, and the bite, as in this case, usually goes unnoticed. She went home with instructions to keep the wound clean.

A few days later she noticed a red streak running down her leg from the original red bruise. The doctors said the bite had become infected. A long and tedious treatment with antibiotics followed. One drug would fail; the doctors would try another; and all the while the painful wound worsened, its center seeping fluid like a weeping eye. The area of dead tissue was growing, and as it grew, new infections set in. The doctors eventually sent the woman to a plastic surgeon.

The surgeon cut a necrotic mass the size of a ripe plum from her calf. The stitches stayed in for another two months. The spider bite had impinged on, and sometimes dominated, her life for almost half a year. It left her with a patchwork scar.

———

Call her experience a midpoint: people have better and worse times with the recluse's necrotic lesions.

A friend of mine camped in a log cabin on the North Platte River in Wyoming. The paneling job in the cabin was shoddy; as my friend learned later in researching his problem, loose paneling is ideal habitat for recluses. Their habit of hiding in crevices is probably the origin of the name *recluse,* though their predatory habits don't mark them as shy. My friend woke one morning with a sore at the inner crease of his elbow. It grew to a painful pimple with a discolored penumbra the size of a dime, from which red streaks radiated. The sore tissue slowly withered, and eventually a little plug of flesh fell out of it, leaving a scar. The healing took six or eight weeks.

This man's uncle, who was bitten decades earlier, wasn't as lucky. No one in his family remembers the circumstances of the bite. They only remember that the lower part of his bitten leg "dissolved."

In another case, an active and powerful three-hundred-pound man was the victim. His friends found it amusing that this Goliath lay in bed whimpering with pain because of a tiny spider. They were less amused after he showed them the rotting portion of his leg.

The flesh affected by a recluse's necrosis never heals. Somehow, the venom turns off the immune system and the body's capacity for repairing itself in that patch of flesh. The victim can only hope the dead area stays small. But sometimes it doesn't.

One woman who made a photograph of her injury available to researchers had a grapefruit-sized chunk of black flesh in her thigh. She eventually survived, but her ordeal left her unable to walk without a cane. Another woman had a basketball's diameter of flesh trimmed from her upper torso to save her life. A few people have died of gangrene following a bite; others have submitted to surgical disfigurement or amputation.

———

A young woman found herself constantly weary and nauseous. The illness lingered for months. She also kept breaking out in odd sores that healed slowly—purple dimples ringed with raised yellow flesh. It was the number of sores that fooled her doctor. She broke out with a new one every few days, and he couldn't explain why. He could only offer speculations about allergies. He suggested she look through her apartment carefully for possible allergens.

After she had looked for the obvious, she and her friends moved all the furniture, hoping to find some hidden factor. When they tried to dismantle the waterbed, they had their answer. Beneath the mattress were dozens of recluses.

Syndrome 2: Immunity

The white rat, that symbol of laboratory experimentation, is useless to toxicologists researching the venom of recluses. Rats do not react to recluse venom. It's hard to study the effects of the venom on people, since people so rarely notice the bite while it's happening. The recluse's fangs pinch in toward the middle, gathering the upper layer of skin in folds and injecting venom between layers, causing very little mechanical injury. But apparently most people share the rat's good fortune: they don't react to a recluse bite at all.

Syndrome 3: Sudden Death

Mice are no good to recluse researchers either. A bitten mouse reacts with a precipitous drop in blood pressure and dies before any useful data can be collected.

The young woman with the infested waterbed reported nausea, exhaustion, headache, and malaise. These symptoms fall under the heading of "systemic reaction," the doctors' term for anything that's not "local." The young woman's mild systemic symptoms are common. But the systemic symptoms of a recluse's bite don't have to be mild.

A man was carrying wood. He put down an armload and noticed a spider hooked fangs-first into the skin of his arm. He caught the spider in a jar and took it with him to the emergency room, where he dropped dead. The spider was a recluse.

Syndrome 4: A Failure of Immunity

I met a woman who was always sick. Every flu bug that hit town gave her at least one round of illness. She never felt right, even when she had no outward symptoms of a sickness. She said she had been in normal health until fifteen years before, when she was bitten.

The bite of the recluse had cost her some tissue from her leg, but she had recovered with only the usual scar. Her tendency toward sickness developed slowly after that. It was years before she faced the fact that something was deeply wrong with her, that her numerous bouts with common illnesses betrayed some fundamental dysfunction. That's when she drew a chart of her medical history, a chart on which one event she'd always considered insignificant now stood out. The spider bite that had given her a few weeks' trouble and a scar stood at the beginning of her decline.

I listened politely to her story and took notes. I had no reason to doubt her facts, but I didn't believe the spider bite could have anything to do with everything else she'd been through. Privately, I suspected she had some undiagnosed disease of the immune system, like lupus. I even entertained the notion that she was a hypochondriac.

The two of us went to hear a lecture given by a toxicologist who specialized in recluse venom. His response to her story boiled down to two main ideas. One: he had heard of such long-term reactions to recluse venom before. Specialists had known about them for several years, and plenty of anecdotal evidence supported the connection. Two: it's extraordinarily difficult to research such a slowly developing, amorphous phenomenon, so nobody knows why it happens or how to cure it. Despite the bad news, the woman left smiling in her vindication.

The toxicologist told me strange things about what the venom does. In some people it convinces the white blood cells to turn traitor and attack the body that made them. That's what causes necrosis. But the venom's interaction with a human immune system is an intricate dance whose implications are breathtaking and terrifying. If you could turn the body's defenses on selectively, as the venom does, you might know more about AIDS and all the other unravelers of the immune system. You could cure cancers without a scalpel or radiation or health-threatening chemicals. You could make white blood cells devour tumors of the brain.

For now, though, we understand almost nothing about the venom and its attendant array of human suffering.

———

The necrosis-causing toxin serves no known purpose for the recluse itself. In this respect, the recluse resembles

the unnecessarily virulent black widow. But the recluse's case is even stranger.

Most spiders have a venom component that paralyzes
or kills insect prey, and the recluse is no exception. But
the chemical in recluse venom that hurts people isn't the
same one that debilitates insects. The black widow's
toxin has an obvious use in catching prey; it just so
happens that this toxin can affect large animals. The
recluse is dangerous because of an extra, apparently
functionless, toxin in its venom. The recluse venom
doesn't even work the way a rattlesnake's does, serving
to make good on threats delivered by a warning system.
The recluse gives no warnings.

————

The recluse spiders became notorious in the United
States in 1957. They didn't arrive here in that year; they
had always been around. People had been bitten, some
badly injured or killed, by recluses. But, as with the
black widow, doctors and scientists didn't believe such a
danger existed, and that fact kept the general public
uneducated, compounding the danger. There's a big
difference in the histories of human interactions with
the two spiders, however, and in the recluses' case the
blame can't be laid entirely on incredulous scientists.
The other factors include the recluses' habits, its looks,
and the cold facts about plumbing.

Before the United States became a country, European

settlers here had gathered some information about the black widow. Many of the Native American tribes knew the gleaming spiders as dangerous, and some tribe members shared that lore with Europeans. I don't know whether any of the tribes had lore about the recluses, but if they did, it doesn't seem to have been recorded by Europeans. There was no tradition of fearing the recluse.

This is where the looks of the two spiders play a part. The widows are distinctive. Their sleek, sematic coloring makes them easy to recognize. The recluses come in several shades of dull. There's a tan species, and several brown ones, and some brown and tan ones, and so on. They don't have distinctive marks. Several recluse species have a violin shape on the carapace, but that mark isn't distinctive. It appears on the back of a certain wolf spider, and a harmless running spider—that's only to list two spiders I've seen sporting it in my own house. Other spiders also share the mark. It's no more distinctive than the name *Smith*.

Another barrier to identifying the recluse as a danger was its initially painless bite. People weren't likely to associate the right kind of spider with the bite unless they saw it in the act. Add to this the venom's unpredictability. Even though it kills some people, the majority have no symptoms, and a majority of those who do have symptoms notice a necrosis instead of systemic effects. Other venomous animals, like the rattlesnake and the black widow, cause varying but far more predictable symptoms.

All these factors kept the general public from developing much knowledge about recluses. That doesn't mean nobody knew about them; an occasional person had a nasty experience with a bite and made a sharp observation. Cases like that go back at least to the 1870s in the United States. At the time, however, scientists and doctors greeted reports of deadly spiders with irritable lectures on the gullibility of the public. There is actually nothing your average scientist hates more than information from nonscientists, all of whom he assumes to be unwashed, idol-worshipping degenerates good only for working on cars. The thing your average scientist despises second most is a fact that doesn't fit his theory—an odd position for somebody who supposedly works from empirical data. But most scientists are human.

Until the early part of the twentieth century, the experts smugly proclaimed there was no such thing as a seriously toxic spider in the United States. Proof of the black widow's killing potential then led most experts to go around saying to impressionable journalists that there was only one poisonous spider in the United States.

In 1934, the same year Dr. Blair published his account of the widow bite that nearly killed him, a scientist named Machiavello published the hard facts on a critter from Chile, the corner spider. This spider hides in clothes and sheets and bites people when they unknowingly squash it. The symptoms it produces include necrotic lesions.

The corner spider is a member of the genus *Loxoceles,* the gang I've been calling the "recluse spiders." In other words, certain spiders in the United States were occasionally getting accused of biting people and causing spots of necrosis; a closely related spider in South America had been proved to cause exactly such spots; and most American scientists persisted in believing the only dangerous spider in the United States was the black widow. The widow got blamed for a lot of recluse bites, and necrotic lesions were considered a symptom of widow bite. (The widow can, in fact, introduce infections that rot the skin as a recluse bite sometimes does.)

Of course, some scientists weren't hog-tied by such prejudices. Some of them documented a few spiders that can cause some mild systemic symptoms in humans. But the biggest influence on this kind of research was indoor plumbing. Widow bites occurred in outhouses far more often than anywhere else. In the 1950s, indoor plumbing was rapidly replacing outdoor facilities across the country, and the number of widow bites dropped dramatically. When it did, doctors noticed how many of the spider bite cases they treated were of the kind that had once been a minority, the kind in which the patient suffered a necrosis. Why should the usual sweat-and-pain widow cases vanish while the odd skin-rot cases persisted? It was as if an ocean had receded, leaving a previously submerged boulder prominent on the beach. Science was ready to believe some other spider capable of injuring humans. In 1957, proof of necrotic symptoms caused by recluse bites reached medical journals, and the information quickly spread to newspapers, magazines, and popular books.

When recluses first received a lot of press, they were often described in ridiculously inaccurate terms. One book referred to them as "large, hairy spiders." One of those three words is correct. I imagine that mistake arose because of the large, hairy wolf spiders that wear fiddlelike marks on their backs. It's even possible to find books from that period that refer to the recluse as a member of the genus *Latrodectus*. But *Latrodectus* only includes the widows, which are as dissimilar from recluses as humans are from giraffes.

For the last forty years the popular press has been spouting science's revised company line: There are only two dangerous spiders in the United States. As usual, the assumption that we know everything has got us into trouble. In 1996 *JAMA: The Journal of the American Medical Association* reported that another spider has caused necrosis and death in humans. The spider's scientific name is *Tegenaria agrestis,* and its common names include *hobo spider* and the mellifluous *aggressive house spider.* The species was described long ago. Its genus, in fact, is found around the world. It simply wasn't recognized as dangerous because everyone knew there was only one dangerous spider around—and later, make that two dangerous spiders. Symptoms of the hobo's bite were often called the work of the recluse, just as the widow long took the blame for the recluse's handiwork.

The truth is, nobody knows how many kinds of dangerous spiders exist. There are a number of proven human-killing spiders around the world besides the ones known in the United States—the Australian trap-door spider, a certain Brazilian wandering spider, a few tropical tarantulas. The diversity of these spiders shows a wild venom can crop up anywhere in the spider family tree. It shouldn't surprise anybody if we discover another deadly one in another forty years.

At a recent conference, one scientist had just come back from the forests of Central America, and he was showing off specimens he had collected. He had dozens of species from the recluse genus, none of them ever described before.

————

I used to come to the shed in the summer to hunt vermin. In the northeast corner I could usually find the web of a plump female black widow spider stretched between the sheet metal walls and the welded steel studs. I would toss a cricket into the web to draw the widow out. She would come cautiously from her hiding place, tapping a long foreleg in front of her, and when the cricket made a few kicks that confirmed her diagnosis of food in the web, she would come for him in rushes separated by pauses when she would listen with her legs. As she moved, her abdomen shivered like a soap bubble near bursting. She moved belly-up, the hourglass on her abdomen red as a double wound in the dim light. Once she was out in the open, I could capture her.

I suppose something about the placement and shape of the building drew the wind into that corner. The young widows must have come in riding the wind every year, for I captured and removed new adults each summer.

It was always easy to find bait for them. Generally a few crickets would be popping around on the cement floor, disturbed by the flood of light when I opened the door. There were pill bugs crawling the floor along the walls (drained shells of pill bugs hung in the lower reaches of the widow web like dirty tassels on a shawl). I could always scare up a few more pill bugs by pouring water into the fissure in the cement. There would be a trail of ants entering the shed on one side and exiting on another, carrying nothing visible and going somewhere for no reason discernible to me. Occasionally a thumb-sized cockroach would race across a cardboard box, its burst of speed beginning with a sound like a striking match, and after it had passed from sight its hiero-glyphic tracks remained in the dust.

A few mud dauber nests clung flat to the crossbeams. They were made of pale mud dried harder than brick (once I tried to crush one with a brick, and the brick yielded first). Eye-high on the sheet metal I might notice the egg case of a mantid—a hard, ridged structure shaped like a burial mound.

The place was crawling with life, but the situation was about to change.

———

The shed needed to be organized. I hadn't been inside it for several years.

Wasp nests still clung to the crossbeams, but no wasps stirred the dusty air. When I tapped a nest, it crumbled. On another nest I found a wisp of spiderweb. There were no crickets visible, and none singing out of sight. There were no roaches, no ants, no pill bugs, no egg cases on the walls. The corner where the widows used to make their webs held nothing. Along the walls there were little tangles of spider web, apparently without structure. These were not the work of widows; the texture was wrong.

But I had work to do. I lifted an old box of tax records. Something tickled my fingers.

I dropped the box and turned it over. A dozen small spiders covered the bottom of it, and a few more lay in the spot where the box had been. The spiders were brown. Their legs were stringy, an average adult set spanning an area slightly larger than a quarter. Most of them did not react to my intrusion, but the two that did moved very fast, darting around to a sheltered side of the box.

I brushed frantically at my arms where they had touched the box. I knew the spiders by their movement: they were recluses.

I went out into the sunlight and examined my fingers where I had felt them touched. I couldn't see any bites.

I returned to the shed wearing gloves. Under every shelf, in every drawer, behind every wall stud, in every crack were the brown spiders, most of them dormant in the afternoon heat. They rested in sleeping bags of silk. When I disturbed them they ran away in arcing paths, their stringy legs working in sequence so fast they resembled the guttering of fire in a stiff wind. Everywhere there were wisps of web. I found a few cottony disks of webbing. Inside these lay orange masses of eggs. There were also the spiders' discarded skins, which were hard to distinguish from the spiders themselves except that the skins' legs were flung back straight.

I captured one spider in a jelly jar. There was no point in taking two. When you put two recluses into a jar, you soon have only one. At the kitchen table, where the light was good, I could see the violin-shaped mark on the spider's carapace, a dark-brown pattern against the light-brown background.

The movement of the specimen before me had identified it well enough, but I looked with the magnifying glass anyway. Rimming the head was a crescent of six simple white eyes. There were two in front and two on each side. These were not the large, brilliant eyes of a jumping spider, who spots insects in the air and leaps to catch them as they land. Nor were they the meager eyes of an orb weaver who waits in a snare for prey. They were the eyes of an ambush predator, one that kills whatever happens by. They were the eyes of a recluse.

The recluse hunts by sight, attacking moving objects. You can fool it into grabbing a stick dragged along the floor in front of it. The spider seizes its prey, injects its venom, and then retreats. As the spider waits at a distance, the venom paralyzes the prey. This combination of speed and paralyzing venom explains how the recluse kills larger, stronger predators. When the prey is well paralyzed, the recluse seizes it again and chews a hole in its exoskeleton. It injects digestive acids from its stomach and drinks the liquefying prey.

I recalled the strange lifeless quiet of the shed. My mind settled on the mud dauber wasps I had often seen there: whippet-thin predators that specialize in hunting spiders to feed their hatchlings. A biochemist had told me once about his adventures collecting brown recluses so he could research their toxin. He had seen a recluse

climb into the nest of a mud dauber. The spider backed
out a moment later dragging the wasp, the spider's
fangs still buried in the wasp's face.

———————

On the island of Guam, in the decades following World
War II, the native birds and lizards began to disappear.
Entire species unique to the island went extinct. The
cause of this ecological apocalypse turned out to be the
brown tree snake, an exotic species that somehow established a population on Guam. Probably humans
brought them in accidentally. The brown tree snake
doesn't cause environmental chaos in Southeast Asia
and Australia, where it's native. There other predators
and parasites check its numbers. But in a new place,
freed from its enemies, the brown tree snake made a
sudden, enormous impact.

The little shed wasn't as isolated from outside populations as an island is. Since new arthropods could crawl
into the shed anytime, the "extinctions" weren't permanent. Still, the similarity is more than an analogy; the
same principle of populations interacting in separate
pockets applies in both. Instead of a brown snake, a
brown spider had brought on this miniature apocalypse.
The spider's aggressive predatory style wiped out every
other species. Of course, a predator that exterminates its
prey must pay the consequences.

The following year, nothing lived in the shed.

Soon enough the cockroaches and ants and crickets would be back. Perhaps the crickets would eat the skins the recluses had left behind. The skins were the only thing left now; they lay everywhere like the cast-off coats of untidy children. If I pried one loose from the web-strands that anchored it to a wall stud or the underside of a shelf, a breeze too light to be otherwise noticed would send it scudding across the floor. I opened a toolbox and saw dozens of skins shivering like palsied hands.

Here's what must have happened over the years: A single recluse wandered into the shed and laid her disk of eggs. They hatched, a brood of indiscriminate predators ready to eat anything small enough, and as they grew larger, so did their prey. The recluse will seize prey items many times its own size. It will eat its own young and its sexual partners.

The recluses reproduced slowly. A recluse needs perhaps five years to grow to its full size—a very slow rate for spiders and other small creatures. The black widow, for example, reaches maturity in about three months, and rarely lives past two years. The fast-running recluse lives a long, slow life. It winters in crevices, sealing itself from the cold with a pocket of webbing. Put it in a jar without food or water and its slow metabolism will sometimes last months.

Eventually every creature that could possibly be eaten by a recluse was extinct in the little world of the shed— every creature except the recluses, which now num- bered at least in the thousands. They had never shied from spilling kindred blood, but now they became sys- tematic cannibals, eating nothing but their own siblings and cousins. Finally there were only a handful left, and they crawled out to find fresher hunting grounds.

———

Recluses regularly take over abandoned buildings this way. Their life span, long for an arthropod, causes their generations to overlap, creating a gradually building population. The takeover typically requires twenty-five years in a reasonably solid building. In a poorly sealed building, the takeover can occur in five years.

A more typical spider species has an annual cycle. Most individuals of such a species will hatch, reproduce, and die within a year. The size of next year's population will depend on this year's crop, a few variable factors like weather, and the species's rates of reproduction and survival, both of which are influenced by many other factors in the environment. It's immensely complicated, but things get even harder when you're dealing with the recluse, whose cycle naturally works in booms and busts. And these changes in population aren't general across a geographic region—they're going on constantly in independent little pockets, like abandoned buildings.

These irregular cycles of conquest, overpopulation, and self-extermination aren't side effects of living with humans. In the wild, recluses take over cliffs of flint or limestone. You can scrape a knife into the crevices of such stone and scare out dozens of recluses in a few minutes. In the desert, their numbers balloon inside the dried skeletons of plants. I remember glancing at a stack of firewood one time and seeing an odd hint of motion, a slight shift of texture that let me know some abundance of arthropods crawled there. Looking closer at the relief map of the bark, I saw nothing; then, after a moment, the arcing trajectory of a recluse's run; then several. I kicked over a chunk of wood and exposed scores of them.

———

Ballooning into a cannibalistic frenzy may seem a strange way to control population, but the phenomenon has parallels in the behavior of diverse animals. Scientists have long known that rats, when crowded into spaces too small for their number, begin to rape and murder their fellows. Their social structure breaks down; mothers stop protecting their young; males betray their allies. Something similar happens in rabbit plagues, the population explosions that often follow such human shenanigans as the mass slaughter of predators. The rabbits raze all vegetation, then chew the limbs off each other.

Most revelatory are the several species of grasshopper called locusts. Usually the individual hoppers reach adulthood and live solitary lives, stuffing their gullets with as much food as they can find before mating. But when they find themselves surrounded by fellow grass-hoppers—a circumstance that comes about when the weather's right for overpopulation—the individuals physically transform. Their color alters; their anatomy shifts. They've become that pestilence known in old times as a plague of locusts. They migrate for hundreds of miles as a swarm, enabling themselves to survive even though their numbers divest the land of vegeta-tion. The swarming state is like an extra phase in the life cycle, one that's activated only if conditions warrant.

Scientists have tended to see behavior as either nor-mal or abnormal—an unstated assumption that has slowed our understanding of aggregate behavior. Many animals have different sets of behaviors for different population densities and food supplies. For the locust, the solitary life and swarming are *both* "normal"—and adaptive. The recluse's population booms and cannibal-istic die-offs occur so consistently they can't be consid-ered "abnormal" either.

The same is true for mammals. That's an uncomfort-able idea, because scientists like to draw parallels be-tween rat and human behavior in high-density populations. If intraspecies violence is in some sense "normal," we have to reorder our ideas about human nature.

Serial murder, war, genocide, and even witch hunts have all been linked to population changes and competition for resources. We let ourselves off the hook when we define such killing as "abnormal." We put the behavior at a distance, letting ourselves think of it as something alien, something we normal folk could never do. We think of the Nazis who murdered millions in death camps as demons, instead of people who faced choices like ours. But the capacity to murder, to become demonic, is in our nature.

One of our natures, anyway.

————

I returned to the shed this spring. It had been three years since I found it in the lifeless aftermath of the brown recluse population explosion. I was shocked to find the building repopulated in predators.

I spotted a fresh-looking black widow egg sac. I used a stick to pull the sac free of its web, which ripped like a rusty zipper, and as I did I disturbed several baby widows that had been feeding on a carcass. When the knobby little widows had cleared away, I recognized the carcass as a brown recluse. As I reached for a jar to collect some of the widows, a big recluse charged over the horizon of the box the jars were in and stood there as if daring me. I backed him down with the stick.

As I searched the walls of the place, I found about twenty widow webs, most of them weighted with last year's dust. Half of them contained swaddled remains I recognized as recluses. I also found enough of the recluses' softer silk to suggest a population in the hundreds. Stuck to a two-by-four with recluse silk was a male widow. Several other bits of exoskeleton littering the walls looked like widows killed by recluses as well. This accounting includes only the remains I picked up. I left a great many patches of wall alone, because I saw enough live examples of both types of spiders to give me a healthy dose of caution and a serious case of the willies.

Besides the spiders, I also found a mud dauber's nest, probably from the previous year. The cells were packed with exoskeletons, which were all that remained of the recluses that had been devoured from the inside by larval wasps. I didn't see any black widow remains in the dauber nest, though the wasps are known to prey on widows.

I also found fresh mantid egg cases and crickets and leafhoppers and some tiny flying insects I couldn't get a good look at. The predators far outnumbered the vegetarians.